INVERSE AND ILL-POSED PROBLEMS SERIES

Theory of Linear Optimization

Also available in the Inverse and Ill-Posed Problems Series:

INVERSE AND ILL-POSED PROBLEMS SERIES

Theory of Linear Optimization

I.I. Eremin

UTRECHT • BOSTON • KÖLN • TOKYO
2002

VSP   Tel: +31 30 692 5790
P.O. Box 346   Fax: +31 30 693 2081
3700 AH Zeist   vsppub@compuserve.com
The Netherlands   www.vsppub.com

© VSP 2002

First published in 2002

ISBN 90-6764-353-X

Printed in The Netherlands by Ridderprint bv, Ridderkerk.

Preface

This monograph is devoted to the basic component of the theory of linear optimisation problems: systems of linear inequalities. Such an approach is exact in both a historical and methodological sense.

The first two chapters are based on two chapters from the monograph (Eremin and Astaf'ev, 1976), the material of which has been transformed, improved and expanded, whereby attention has been focused on economic interpretation of models, theorems, and approaches.

The other chapters are dedicated to less traditional problems of linear optimisation, such as: contradictory problems and duality, lexicographic problems and duality, piecewise linear problems and duality, etc. Duality problems are thus treated throughout the monograph.

The book also covers some general methods for calculating processes for certain problems of linear optimisation: the problem of stability and correctness.

This book contains original scientific material which is useful to students and specialists in mathematical optimisation, operation research, economic-mathematical modelling and related disciplines, and could also be used as a textbook.

The general mathematical basis of linear optimisation is linear algebra. All necessary facts in linear algebra can be found, for example, in (Mal'tsev, 1975).

The applications given at the end of the book serve a twofold purpose. Firstly they give information on linear optimisation relatively non-linear (convex) programming in a brief canonical form. Secondly, it allows the reader to refer to the facts of non-linear optimisation to justify certain theorems.

The bibliography contains mostly books in mathematical programming and consists of educational and scientific books in Russian (some of which have been translated).

The numbering of paragraphs is open; the numbering of theorems, lemmas and remarks has two indices: the first index is the number of the paragraph, the second index is the number of the theorem (lemma, remark).

The author wishes to thank his colleagues at the Department of Mathematical Programming of the Institute of Mathematics and Mechanics Ur. Otdel. Ros. Akad. Nauk: N.N. Astaf'ev, L.D. Popov, V.D. Skarin, and S.V. Plotnikov for reading the text of the book and their useful remarks. The author also wishes to thank G.F. Kornilova for compiling and editing the text of the book.

This edition is partly supported by the Russian Fund of Fundamental Research (N-98-01-14016).

Contents

Introduction

The problems of linear optimization are the extremal problems which are formed by linear functions and linear relations. In any case the basis problem here is the problem of linear programming (LP problem). This is the problem of extremum search (maximum or minimum) of a linear function with the restrictions in the form of linear inequalities. A great importance of such problems is defined by their economic contents (30–50 years of 20th century). Further the interest had not weakened.

The LP problem (in the narrow sense) is the problem of maximization (or minimization) of a linear function with restrictions in the form of a finite system of linear inequalities, for example:

$$L : \max\{(c, x) \mid Ax \le b\}, \tag{1}$$

where $A = [a_{ij}]_{1,1}^{m,n}$, $c = [c_1, \ldots, c_n]^T \in \mathbb{E}_n$, $x = [x_1, \ldots, x_n]^T \in \mathbb{E}_n$, $b = [b_1, \ldots, b_m]^T \in \mathbb{E}_m$. In the general case the system of restrictions may contain the inequalities both "$\le$" and "$\ge$" and the equations. Nevertheless, form (1) for LP problems is one of standard forms.

The forming subobject of problem (1) is the system of linear inequalities

$$Ax \sim b \sim (a_j, x) \le b_j, \quad j = 1, \ldots, m, \tag{2}$$

where a_j is j-th row of the matrix A, $(\cdot, \cdot)$ is the scalar product. System (2) sets *the polyhedron* $M := \{x \mid Ax \le b\} \subset \mathbb{E}_n$. If $\max_{x \in M}(c, x) =: \alpha < +\infty$ then the problem L is to find

$$\bar{x} \in M \cap \{x \mid (c, x) = \alpha\} \quad (=: \operatorname{Arg} L).$$

If we add the restriction of non-negativeness of $x \ge 0$, then the problem L has a simple economic interpretation. Suppose $x_i \ge 0$ is the quantity of production of i-product; c_i is the realization price of the unit of i-product;

b_j is the quantity of j ingredient. These are the production–forming factors. i-th column of A is the vector of ingredient expenditure on the unit of j-product. Then x is the plan of production; (a_j, x) is the expenditure of j-th ingredient for the plan x; (c, x) is the income defined by the vector x. The model L is *the problem of search of the production plan with given resource restrictions which provides the maximal income.*

LP problem admits numerous generalization connected with generalizations both of the initial space $\mathbb{X}$ (in (1) it is $\mathbb{E}_n$) and of the number of restrictions

$$\max\big\{c(x) \mid a_\alpha(x) \le b_\alpha, \ \alpha \in \Omega\big\}, \tag{3}$$

where $\{c, a_\alpha \mid \alpha \in \Omega\}$ are the elements of the conjugate space $\mathbb{X}^*$; Ω is the set of indices of an arbitrary structure (for example, Ω may be a continuum). Model (3) describes a wide class of problems from various sections of mathematics (theory of function approximations, theory of optimal control and so on). Another generalization of the problem L is the problem of multicriteria optimization, when the admissible $x \in M$ is estimated by several criteria. Refinements of these setting are connected with Pareto optimization, lexicographic optimization, and *mixed optimization* – Pareto – lexicographic. In the limits of linear programming we select integer programming, when, parallel with ordinary restrictions (inequalities and equalities), we require that some coordinates of x (or all coordinates) are integers. This restriction makes the problem essentially more complicated and requires to use combinatorial analysis.

Linear programming, formally, may be considered as specific branch of theory of linear inequalities, which is the basis element for LP. All essential moments in LP are corollaries from theorems for systems of linear inequalities. These facts are: the Farkas–Minkowski theorem on dependent inequalities, the Chernikov principle of boundary solutions; dual representation of polyhedrons; analytic conditions of polyhedron boundedness, stability conditions and so on. One of creators of LP Dantsig, D. in Introduction to his monograph "Linear programming and extension" (1963), (Dantsig, 1966) says that the book "devoted to systems of linear inequalities and their solution". The basic theorem of theory of linear inequalities is the Farkas theorem (Farkas, J. Theorie der einfachen Ungleichungen. J. Reine Angew. Math. 1901, Bd. 124, pp. 1–27). In our monograph this is the Farkas-Minkovski theorem. This theorem in one of formulations is as follows: if the inequality $(c, x) \le \alpha$ is the corollary of the system $Ax \le b$, then there exist non-negative numbers $\{\bar{u}_j\}_0^m$ such that $(c, x) - \alpha \overset{(x)}{\equiv} (\bar{u}, Ax - b) - \bar{u}_0$. In this case, if $\{x \mid (c, x) - \alpha = 0\} \cap M \ne \varnothing$ then $\bar{u}_0 = 0$ and $\alpha = \max_{Ax \le b}(c, x)$.

This theorem is the basic instrument for forming and justification of duality in *LP*. In particular, this allows to:

1) realize constructive reduction of *LP* problem to solution of a certain system of linear inequalities given explicitly.

2) calculate derivatives of the optimum function $f(A, b, c)$ of the *LP* problem using the problem parameters. In general case this allows to calculate the directional derivative in the space of all the parameters (in (1) these are the elements which form the matrix A, vectors b and c).

3) realize reduction of the matrix game in mixed strategies to the pair of mutually dual *LP* problems; therefore, to a certain system of linear inequalities.

We note also that duality in *LP* has deep economic contents. This denotes that any mathematical–economic theory that we can construct is less effective without use of the duality in *LP*.

Chernikov, S. N. (1968) in Introduction characterizes the stakes in developing theory of linear inequalities mentioning such names as Fourier (1823), Farcas (1894), Minkovski (1896), Voronoy (1908), Stimke (1915), Haar (1924), Dains (1933), Veil (1935), Motskin (1954) and some others. The first work of Chernikov, S. N. on linear inequalities was done in 1944. The expanded paper "Systems of linear inequalities" was published by Chernikov, S. N. in j. Uspekhi Mat. Nauk, 1953, 8, 2(54), pp. 7–73. The contribution of Chernikov S. N. in development of theory of linear inequalities is rather essential: he had formulated the principle of knot solutions; he had developed the Fourier method and the algebraic theory of infinite systems of linear inequalities with finiteness conditions. His monograph (Chernikov, S. N., 1968) is up to now the most global source in theory of linear inequalities in russian literature.

The development of *LP* appreciably had the evolutionary character, however, the jerks in 30th, 40th, and 50th become possible since these problems were arised from economic practice. These were the problems of control by complicated integrated systems of war–economic contents, the problems of operation research (before war and during the war period).

Now, we touch the history of linear programming. Even Fourier had considered the systems of linear inequalities. He had suggested for such systems the method is known as the Fourier method. Fourier even solved the extremal linear problems with restrictions with few number of unknowns. Nowever, his efforts in the solutions methods were not completed.

The essential contribution even the break in development of linear programming, in modelling and solution of applied economic problems was the paper of Kantorovich, L. V. (1939). In his further work "Mathematical problems of optimal planning" (in "Math. Models and Methods of Optimal Planning", Nauka, Novosibirsk, 1966) he will say that we need to use a lot of mathematicians from various sections of mathematics to develop this direction. The paper of Kantorovich L. V. (1939) had 9 sections each of which was devoted to a certain economic problems. Also it had contained three applications dedicated to the method suggested by Kantorovich to solve these problem – the method of *dividing multipliers*. This method was the pre–image of the simplex–method developed by Dantsig and the duality in *LP* which become completed by efforts of Dantsig and Neumann (von Neumann, Dj., and Mongenshtern, O., 1970). We shall mention this monograph further Kantorovich (1939) had considered, in particular, the problems: distribution of details by lathes, maximal diminishing the waste products, the best partitioning the area under crops, the best plan of transportations and others.

In 30th several works devoted to transport models of linear programming and methods of their solving were published by the economist from Leningrad Tolstoy A. N. (for example, Methods of elimination of nonrational transportations in planning in Social transport, 1939, 9, pp. 28–51).

The end of 40th and the beginning of 50th was the period of general interest to linear programming, essentially in USA. Therefore, parallelism, the independent obtaining the results were rather usual. This had noted, in particular, by Dantsig Dj., in his "Memoirs on arising the linear programming" (in Memoirs of the American Math Society, 1984, 48, N298). Thus, the enumeration of names who had established the important results in *LP* was not simple.

As for *LP* in our country, we have to note forming of scientific directions by optimization problems (Kantorovich, Chernikov, Kiev School of Glushkov and others). This direction of *LP* was influenced by appearance of the translated books of Kun, G. U. and Takker, A. U. (1959) and Errow, K. Dj., Gurvitz, L., and Udzava, H. (1962).

We want to mention also the two publications on history of linear and nonlinear programming which show us how this direction was formed. This is the mentioned essay of Dantsig, Dj. (1984) and the paper of Prekopa, A. (1980). On the development of Optimization theory. In: Amer. Math. Monthly, 1980, 87, 7: pp. 527–542.

The paper of Prekopa is devoted basically to those sources which are connected with research of mechanicians in equilibrium and stability in mechanic systems fulfilled in the last century and earlier (Fourier, Lagrange, Kurno, Farkas, Gauss, Ostrogradskii, Gammel and so on). In 1788, Lagrange J. L. in his book "Analytic mechanics" had considered the problem

$$\min\{f_0(x) \mid f_j(x) = 0, \ j \in J\} \tag{4}$$

as the model of search of equilibrium state of mechanical system which is defined via the minimum of the function. $f_0(x)$ of kinetic energy (Lagrange, J. L., 1788, Mecanique Analytique 1-2, Paris). Lagrange had written the necessary equilibrium conditions via the function (the Lagrange function):

$$F(x, u) = f_0(x) - \sum_J u_j f_j(x)$$

in the form

$$\nabla_u F(x, u) = 0, \qquad \nabla_x F(x, u) = 0.$$

This means that the state vector x is admissible and satisfies the equation

$$\nabla f_0(x) = \sum_J u_j \nabla f_j(x) \tag{5}$$

for certain u. Fourier (1798) had considered the case when in the system of relations there were inequalities, suppose $g_i(x) \geq 0$, $i \in I$. Analog of conditions (5), in this case, attains the form (Kurno, 1827; Ostrogradskii, 1934; Farkas, 1894)

$$\nabla f_0(x) = \sum_J u_j \nabla f_j(x) + \sum_{I(x)} v_i \nabla g_i(x), \quad v_i \geq 0 \tag{6}$$

here $I(x) = \{\, i \mid g_i(x) = 0 \,\}$.

The linearized variant of the problem in question (or initially linear), i.e.

$$\min\{(c, x) \mid Bx = d, \ Ax \geq b\}, \tag{7}$$

transforms relation (6) into relations

$$c = B^T u + A^T v, \quad v \geq 0 \qquad and \qquad (Ax - b, v) = 0. \tag{8}$$

These relations contain in itself the duality theorem in *LP* in its modern setting. Duality in linear programming can be obtained from the Farkas theorem. Really, if $\alpha := \text{opt}\,(7)$, then the inequality $(c, x) \geq \alpha$ is the corollary of system of restrictions of problem (7). Then, by the Farkas theorem, we have

$$(c, x) - \alpha \overset{(x)}{\cong} (\bar{u}, Bx - d) + (\bar{v}, Ax - b)$$

for certain $\bar{u}$ and $\bar{v} \geq 0$. From this identity follow conditions (8).

Of course, it is the simplified variant of setting the evolution of the problem in 19th century and in the beginning of 20th century. The necessary facts in their rigorous form were obtained slowly, at first, with errors. Really, forming the notion was more hard than, for example, proving an important mathematical theorem. As for a new scientific direction – this way is rather more hard.

In "Memoirs ...." of Dantsig the situation of initial forming LP in USA was as follows. The time which Dantsig described is 1950–1970, partially, 1970–1980. Dantsig, at first, had noted that LP was the revolutionary achievement which gives the possibility to formulate the general goal and to find by means of the simplex–method the optimal solutions for complicated practical problems. He marks 1947 year as the year when LP had arised (in connection with plan work of military). Note that the school of Kantorovich, L. V. marks the year 1939 as the year of similar sense. As for Dantsig, he had mentioned the contribution of Kantarovich by two phrases: "the important research being done by Kantorovich in 1939 wan not noticed in Soviet Union. Only after mathematical programming began to develop in the East (1959) the work of Kantorovich had become known."

Dantsig connects his own interest and contribution with the experience of solution of program planning with the use of calculation technology of the second world war period. One of the problems was related to the Air Force Department (1946). Basing on their model with different branches of Leont'ev "expenditure–output" (1932) Dantsig had generalized it. He introduced the notion of technology processes, their intensities, products, goal setting and so on.

The model that Dantsig had obtained was of large dimension (because of the dynamic setting) and the known methods had failed. The exhaustive search also were not effective because of a large number of variants (for the assignment problem 70×70 the number of variants is $70! > 10^{100}$). The search of the way of overcoming the difficulties had led Dantsig to economist Kupmans T. K. and his young colleagues from Kauls Fund from Chicago University. Kupmans had solved the problem of Combined control of transportations. As a result, Dantsig had suggested in summer, 1947, the simplex–method (close to those that Kantorovich had suggested in 1939).

In october, 1947, Dantsig had met Dj. von Neuman from the Institute of Fundamental Research in Prinston and had discussed the results. It had occured that in the framework of his research, in the game theory von Neuman had known about theory of linear programming and duality. By then, the book of von Neuman and O. Morgenshtern "Game theory and economic behavior" was finished. In this book, in particular, was shown the equivalence of two mathematical situations: linear programming and matrix games in mixed strategies.

The research of the Prinston group: Takker, A. and his learners Kun, G. and Geil, D. on game theory, nonlinear programming, and duality theory may be related to 1948. Dantsig in 1948 had met Takker. As a result, we have three sources of duality theorem: von Neuman, Dantsig and Takker group. All these results are related to a short period.

In 1949, in Chicago University the first conference on mathematical programming had taken place. The following economists and mathematicians had participated (as Dantsig had noted) in this conference: Kupmans, Erron, Gurvits, Samuelson, Dorfman, Georgesku–Rogen, Simon, Takker, Kun and Geil.

As for the term "linear programming" Dantsig had written that the military men call their plans, praphs and manoeuvres, and so on by *programs*. So, when Dantsig had analyzed the problem of planning in Air Force and saw that it may be formulated in the form of linear inequalities, he had called his first paper (on this theme) as "Programming in a linear structure". In summer, 1948, Dantsig and Kupmans had visited the corporation RAND. One day, they were walking near the sea–coast Santa–Monika. Kupman asked Dantsig whether "programming in linear structure" he may shorten as "linear programming". Dantsig had answered that it was a good idea. So this term had appeared. The term "mathematical programming" was introduced by Dorfman (1949). The term "simplex–method" had arisen as a result of exchange of opinions between Dantsig and Motskin T. (1949).

His paper Dantsig finishes by the following jungement: "If I have to summarize my initial contribution in linear programming, I would describe it as follows:"

(1) Discovery (as a result of working in practical planning during five years) that most of relations in practical planning can be reformulated as systems of linear inequalities.

(2) Expression of criterion of choice of available and the best plans explicitly (for example, as a linear goal function), but now the collectionof

basic principles and laws which, in the best case, may be only the tool to attain the goal.

(3) Investigation of the simplex–method which had converted rather interesting approach to economy theory to the basic instrument of practical planning of large complicated systems.

It should be said in all fairness that Kantorovich, L. V. in 1939 may summarize his contribution by the points (1), (2), and in modified form by (3). Of course, in the end of 40th all that was connected with linear programming was seen deeper. But it is the merit of *time*.

Summarizing all that we say relatively the history of formation of linear (and nonlinear) programming we should note the following stakes:

1. Fourier method for systems of linear inequalities.

2. Optimization Lagrange principle and its generalization on the case of restriction given in inequality form.

3. Farkas theorem on dependent inequalities.

4. The Kantorovich paper (1939).

5. Simplex–method of Dantsig and Neuman duality (the end of 40th).

As for the works of mechanicians on the Lagrange principle (this clarifies Prekopa in the paper "On development of Optimization theory"), in the chain of this stakes one can see the paradox situation: nonlinear programming had arisen at most on 150 years earlier than linear programming. But this is following the internal logic of idea development. As for their external form, they had different appearance and different form, so their depth identity was not clear.

Now we shall consider some aspects of contents of the book. The first two chapters correspond to the first two chapters of Eryemin and Astaf'ev (1976) but they are essentially transformed. In the framework of the monograph they are basic. The apparatus introduced in these chapters is applied in all sequential chapters. Chapter 3 is devoted to improper problems of linear programming ($IPLP$). The basic problems of theory of these problems are stated here: duality theory, methods of optimal correction. In this chapter we had not stated the results on transfer of $IPLP$ theory on the arbitrary ordered spaces (Eryemin, 1988, Chapter 5). We have not included these problems since we want to restrict ourselves by a certain level of proof complexity and the apparatus which we have introduced. As for the general idea of improper (contradictory) problems, we may say the following. In theory of certain classes of problems (mathematical models) we can see weakening of requirements relatively the mathematical object in question. Evolution, in principle, we determined by the influence of the applied problems. We can introduce the following scheme

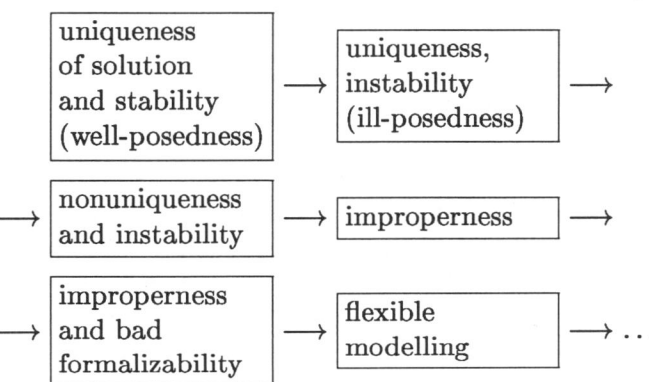

Well developed mathematical methods correspond to first links of this chain. The further motion along this chain leads to greater lack of mathematical (formal) methods of exact analysis of correspondent applied problems (for example, economic problems or the problems of complicated operation medium). In this situation the development of new formalizations, mathematical theories, modelling methods becomes the complicated problem of science progress.

As for the term "bad formalizability" in the scheme above, its clarification on the basis of the apparatus of pattern recognition can be found in Eryemin and Mazurov (1979).

In the limits of theory of linear optimization we have various settings of linear problems, however, which have the identical interior sense. Such problems (except LP problems) are the problems of multicriteria (in particular, Pareto) optimization, lexicographic optimization, matrix games (and their certain generalizations), problems of linear discrimination, pattern recognition and so on. For all these problems the main interesting theoretical problem is their duality. Chapter 4 devoted to lexicographic problems has the main goal – the construction of symmetric duality. The scheme of realization of such duality is comparison of the lexicographic problem L_{lex} with such a problem L_{lex}^*–lexicographically dual, so that this pair of problems had mathematically informative and useful properties. In this case, the rule of comparison (∗) has the property $L_{lex}^* \xrightarrow{(*)} L_{lex}$. Lexicographic duality allows us to reduce the lex–problem to the system of linear inequalities (as in the case of LP problem).

In Chapter 5 we have considered the problems of stability for *LP* problems. Note that this problems of stability are considered the positions of their similar significance both for solvable and unsolvable problems (by the kinds of their unsolvability). This means that if, for example, the initial *LP* problem is of the first kind, then this problem is identified in stability property in the sense of being improper of the first kind for small variations of whole system of initial data.

Chapter 5 is dedicated to iterative methods of projection type for solution of systems of linear inequalities and problems of linear programming. This class of problems is interesting from various points of view; in particular, from the point of view of their use for nonstationary (evolution) modelling. Stability of iterative methods allows us to change the information parameter $I \in \mathbb{R}^N$ of iteration operator $\Phi(x, I)$ which forms from the element x_k the next element $x_{k+1} \in \Phi(x_k, I)$ from iteration to iteration. In this case, the iterative process will be subjected to the relation $x_{k+1} \in \Phi(x_k, I_k)$, where I_k is information terms of the model in the moment k. The moment k may correspond to the time moment t_k if we consider the model evolving in time. In this case we may apply the term "tracking provess" for the process $\{x_{k+1} \in \Phi(x_k, I_k)\}_k$ (Eremin and Mazurov, 1979).

Iteration operators which we mention are related to the class of Fejer operators (Eremin and Mazurov, 1979). As for the methods, their construction, i.e., the construction of correspondent Fejer operators is based on a certain superposition of elementary projections; namely, projections on the half–spaces. The projection on the half–space setted by a linear inequality is the elementary brick from which we construct the procedure of generalization of the sequence $\{x_k\}$ solving the problem of linear inequalities or it is the problem of linear programming.

The contents of the concluding chapter are the problems of the algebra of piecewise linear functions (k-functions) and the optimization problems formed from such functions. For the problems of piecewise linear programming almost all the basic facts of linear programming hold true. This gives us the possibility to apply the mathematical apparatus which is developed in *LP* to the problems of piecewise linear programming (k-problems). The constructive algebra of k-functions gives us the possibility to reduce piecewise linear problems to solution of certain collection of *LP* problems. There arises the possibility of creating standard computer problems of general–logical level provided us to solve the k-problems by the *LP* methods. If we note that many problems of nonlinear optimization are approximated by k-problems, the role of apparatus of k-functions becomes practically certain.

Chapter 1.

Finite systems of linear inequalities

---

1.   BASIC DEFINITIONS

Let $\mathbb{X}$ be a real linear space and $\mathbb{X}^*$ the dual space, i. e., $\mathbb{X}^*$ is the space of all linear functionals defined on $\mathbb{X}$. We shall write a linear functional $h(x) \in \mathbb{X}^*$ as (h, x). Another notation is $h : x \to (h, x)$. In the case of a finite-dimensional space equipped with a scalar product, (h, x) will be understood as the scalar product. The symbol $\mathbb{E}_n$ will denote a real finite-dimensional linear space of dimension n.

The zero element of each of the spaces in question (in particular, of the dual space) will be denoted by one and the same symbol 0.

Elements of the space $\mathbb{X}$ will be called *vectors* or *points* without drawing a distinction between these two terms.

A finite system of linear inequalities over the space $\mathbb{X}$ is a system

$$(a_j, x) \le b_j, \qquad j = 1, \ldots, m, \tag{1.1}$$

where $a_j \in \mathbb{X}^*$, $b_j \in \mathbb{R}$, $j = 1, \ldots, m$. The *rank* r of system (1.1) is the maximal number of linearly independent vectors in the system of vectors $\{a_1, \ldots, a_m\}$. A vector x_0 is called a *solution* of system (1.1) if it satisfies all its inequalities. System (1.1) is called *compatible* if it has at least one solution.

If $(a, x) \le \alpha$ is a linear inequality over $\mathbb{X}$ and $a \neq 0$, then the set

$$P = \{x \in \mathbb{X} \mid (a, x) \le \alpha\}$$

is called the *half-space* of the space $\mathbb{X}$ corresponding to this inequality. The set

$$P^0 = \{x \in \mathbb{X} \mid (a, x) < \alpha\}$$

is called an *open* half-space. The locus of points

$$H = \{x \in \mathbb{X} \mid (a, x) = \alpha\}$$

is called a *hyperplane*.

If $a = 0$, then the inequality $(a, x) \leq b$ is either inconsistent (if b is negative) or identical (if b is nonnegative). In this case we call the inequality *improper*. The half-space corresponding to this inequality is also called *improper*. It is either $\varnothing$ or the whole $\mathbb{X}$, respectively. Similarly we introduce an *improper* hyperplane H (if $b \neq 0$, then $H = \varnothing$; if $b = 0$, then $H = \mathbb{X}$).

A half-space P is a *convex* set. This means that if $x, y \in P$, then P contains also the segment $\overline{x, y}$ connecting the points x and y

$$\overline{x, y} := \{z \in \mathbb{X} \mid z = \alpha x + (1 - \alpha)y, \ 0 \leq \alpha \leq 1\}.$$

If system (1.1) is compatible, then the set of its solutions is the intersection of a finite number of half-spaces and, therefore, is convex.

The nonempty set of all solutions of the system of linear inequalities (1.1) is called the convex *polyhedral* set or the *polyhedron* of its solutions.

Note that many problems arising when considering system (1.1) can be reduced to the case of a finite-dimensional space. The scheme of such a reduction is as follows. If r is the rank of system (1.1), then the dimension of the factor space $\mathbb{X}/E$ of the space $\mathbb{X}$ with respect to the kernel

$$E = \{x \mid (a_j, x) = 0, \ j = 1, \ldots, m\}$$

is equal to r. Hence it follows that $\mathbb{X}$ can be represented as the direct sum

$$\mathbb{X} = \mathbb{X}_r + E$$

of the r-dimensional subspace $\mathbb{X}_r$ (let $e_1, \ldots, e_r$ be its basis) and the kernel E. Thus, each $x \in \mathbb{X}$ can be decomposed uniquely as follows:

$$x = \sum_{i=1}^{r} x_i e_i + y, \qquad x_i \in \mathbb{R}, \quad y \in E.$$

Setting $(a_j, e_i) = a_{ji}$ and substituting the expression for x into (1.1), we obtain the system of inequalities

$$\sum_{i=1}^{r} a_{ji} x_i \leq b_j, \qquad j = 1, \ldots, m. \tag{1.2}$$

There is a simple connection between the sets of solutions of systems (1.1) and (1.2): if x solves system (1.1), then $[x_1, \ldots, x_r]$ solves system (1.2) and conversely, if $[x_1, \ldots, x_r]$ is a solution of system (1.2) then, for arbitrary $y \in E$, the vector

$$x = \sum_{i=1}^{r} x_i e_i + y$$

is a solution of system (1.1).

Evidently, the rank of system (1.2) understood as the maximal number of linearly independent linear forms in the system

$$\left\{ \sum_{i=1}^{r} a_{ji} x_i \right\}_{j=1}^{m}$$

is equal to r. This rank coincides with the rank of the matrix $[a_{ji}]_{m,r}$.

The above reduction of the general system (1.1) to the case of a finite-dimensional space allows us to consider the initial space as finite-dimensional. Without any loss of generality, we may also take $a_j \neq 0$, $j = 1, \ldots, m$.

Further we shall use such notions as a convex cone, a convex hull, and a conic hull. Let us introduce the corresponding definitions.

A set $K \subset \mathbb{X}$ is called a *convex cone* if

$$0 \in K; \qquad x \in K \Rightarrow \lambda x \in K, \quad \forall \lambda \geq 0;$$
$$x, y \in K \Rightarrow x + y \in K.$$

A *convex hull* of a finite set of vectors $\{a_j\}_1^m$ is the set of their convex combination, i. e.,

$$\left\{ \sum_{j=1}^{m} \alpha_j a_j \mid \sum_{j=1}^{m} \alpha_j = 1, \ \alpha_j \geq 0, \ j = 1, \ldots, m \right\} =: \text{co}\{a_j\}_1^m.$$

A *conic hull* of the set $\{a_j\}_1^m$ is the set

$$\left\{ \sum_{j=1}^{m} \alpha_j a_j \mid \alpha_j \geq 0, \ j = 1, \ldots, m \right\} =: \text{cone}\{a_j\}_1^m.$$

The sets $\text{co}\{a_j\}_1^m$ and $\text{cone}\{a_j\}_1^m$ are convex, which can be easily verified.

2.   THE STRUCTURE OF POLYHEDRONS

We shall assume below that $a_j \neq 0$, $j = 1, \ldots, m$. This assumption is not essential for all the statement to be proved below. However, it allows to dispense with various stipulations which would be necessary if some of a_j were zeros.

The system

$$\begin{cases} (a_{j_s}, x) = b_{j_s}, & s = 1, \ldots, k, \\ (a_j, x) \leq b_j, & j \neq j_1, \ldots, j_k, \end{cases} \tag{2.1}$$

will be called a *k-boundary* system for (1.1) if $\{a_{js}\}_1^k$ are linearly independent.

Definition 2.1. If the set of solutions $M(j_1, \ldots, j_k)$ of system (2.1) is not empty, then it is called the *k-face* of the polyhedron M of solutions of system (1.1).

This definition generalizes the notions of a *vertex*, an *edge*, a *face*, etc.

Definition 2.2. An inequality $(a, x) \leq \alpha$ is called an *implication* of a system of linear inequalities (and equations) if an arbitrary solution of the system satisfies this inequality.

Lemma 2.1. *Let a system*

$$\begin{cases} (a, x) \leq \alpha, \\ (a_{j_s}, x) = b_{j_s}, & s = 1, \ldots, k, \end{cases} \tag{2.2}$$

be compatible. The inequality $(a, x) \leq \alpha$ *is an implication of the system of linear equations*

$$(a_{j_s}, x) = b_{j_s}, \quad s = 1, \ldots, k, \tag{2.3}$$

if and only if

$$a = \sum_{s=1}^{k} \lambda_s a_{j_s} \tag{2.4}$$

for some real λ_j, $j = 1, \ldots, k$.

Proof. Without loss of generality we may assume that $\{a_{j_s}\}_1^k$ are linearly independent.

Sufficiency. Suppose that (2.4) holds. Then an arbitrary solution $\bar{x}$ of system (2.3) can be represented as $x_0 + \tilde{x}$, where x_0 is a fixed solution of system (2.2) and $\tilde{x}$ is an arbitrary solution of the homogeneous system $(a_{j_s}, x) = 0$, $s = 1, \ldots, k$. As a result we obtain

$$(a, \bar{x}) - \alpha = \sum_{s=1}^{k} \lambda_s (a_{j_s}, x_0 + \tilde{x}) - \alpha$$

$$= \sum_{s=1}^{k} \lambda_s (a_{j_s}, x_0) - \alpha = (a, x_0) - \alpha \leq 0,$$

i. e., an arbitrary solution $\bar{x}$ of system (2.3) satisfies the inequality $(a, x) \leq \alpha$. Thus the sufficiency is established.

Necessity. We assume that (2.4) does not hold, i. e., a is not a linear combination of $\{a_{j_s}\}_1^k$. Then since the system of vectors $\{a, a_{j_1}, a_{j_2}, \ldots, a_{j_k}\}$ is linearly independent, the system of linear equations

$$\begin{cases} (a, x) = d, \\ (a_{j_s}, x) = b_{j_s}, \quad s = 1, \ldots, k, \end{cases} \tag{2.5}$$

is compatible for each d. Let $d > \alpha$ and let x' be some solution of system (2.5). Then x' satisfies system (2.3); however, $(a, x') = d > \alpha$. So, x' does not satisfy the inequality $(a, x) \leq \alpha$. $\square$

Before formulating a theorem on existence of k-faces, we consider one auxiliary construction.

Suppose that $p \in M$ and $q \notin M$,

$$\overline{p, q} = \{x(t) = (1 - t)p + tq \mid 0 \leq t \leq 1\},$$
$$s(q) = \{j \mid (a_j, q) - b_j > 0\}.$$

The segment $\overline{p, q}$ *leaves* M at the point $x(\bar{t})$ that is determined as follows: if t_j, $j \in s(q)$, are such that

$$(a_j, (1 - \bar{t}_j)p + \bar{t}_j q) = b_j,$$

then we set $\bar{t} = \min_{j \in s(q)} \bar{t}_j$ and obtain the desired point $x(\bar{t})$.

Theorem 2.1. *Let system* (1.1) *be compatible and its rank be* $r > 0$. *Then for each* $0 < k \leq r$ *there exists a* k-*face.*

Proof. The proof is by induction on k. Let $k = 1$. Let j_1 be any of those indices for which the minimum $\bar{t} = \min_{j \in s(q)} \bar{t}_j$ from the above construction is attained, then the point $x(\bar{t})$ solves the 1-boundary system

$$\begin{cases} (a_{j_1}, x) = b_{j_1}, \\ (a_j, x) \le b_j, \quad \forall j \ne j_1. \end{cases}$$

Suppose that the theorem holds for $k < r$, i.e., there exists a compatible k-boundary system (2.1). Let us show that there exist a $(k+1)$-boundary system and, therefore, a $(k+1)$-face.

Let us take some solution p of system (2.1) and a solution q of the system

$$(a_{j_s}, x) = b_{j_s}, \qquad s = 1, \ldots, k,$$

which does not solve system (2.1). We can choose such q, for otherwise, by Lemma 2.1, all a_j, $j \notin \{j_s\}_1^k$, would linearly depend on $\{a_{j_s}\}_{s=1}^k$, which contradicts the assumption $k < r$. The choice of a new index j_{k+1} for formation of the $(k+1)$-boundary system is made according to the above construction but applied to system (2.1). Namely, we take j_{k+1} from the condition $\min_{j \in s(q)} \bar{t}_j = \bar{t}_{j_{k+1}}$. Then, on the one hand, $a_{j_{k+1}}$ is linearly independent of $\{a_{j_s}\}_1^k$, and on the other hand, the vector $x(\bar{t}_{j_{k+1}})$ satisfies the $(k+1)$-boundary system

$$\begin{cases} (a_{j_s}, x) = b_{j_s}, \quad s = 1, \ldots, k+1, \\ (a_j, x) \le b_j, \quad j \ne j_1, \ldots, j_{k+1}, \end{cases}$$

i.e., the system defines a $(k+1)$-face. □

We shall call each r-face of the polyhedron a *minimal* face. If, in (2.1), $k = r$, then the r-face given by system (2.1) coincides with the set of solutions of the system of equations

$$(a_{j_s}, x) = b_{j_s}, \qquad s = 1, \ldots, r.$$

This is so because, by Lemma 2.1, each solution of the latter system satisfies inequalities from (2.1).

We shall call an $(n-1)$-face an *edge* if it is not a point. From geometrical viewpoint an edge may be a segment, a ray, or a line.

If $r = n$, then an n-face is called a *vertex*. The notion of a vertex of the polyhedron M coincides with the notion of its *extreme* point, i.e., a point which cannot be the middle of a nontrivial segment belonging to M. Let us prove this. An n-face, i.e., a vertex, is defined by the system

$$\begin{cases} (a_{j_s}, x) = b_{j_s}, & s = 1, \ldots, n, \\ (a_j, x) \le b_j, & j \ne j_1, \ldots, j_n, \end{cases} \tag{2.6}$$

with a linearly independent system of vectors $\{a_{j_s}\}_1^n$. Let $\bar{x}$ be such a vertex. If $\bar{x} = (p + q)/2$ and $p \ne q$, $p \in M$, $q \in M$, then

$$(a_{j_s}, p + q) = 2b_{j_s}, \qquad s = 1, \ldots, n.$$

But since $(a_{j_s}, p) \le b_{j_s}$ and $(a_{j_s}, q) \le b_{j_s}$, therefore,

$$(a_{j_s}, p) = b_{j_s}, \quad (a_{j_s}, q) = b_{j_s}, \qquad s = 1, \ldots, n.$$

However, the system of equations from (2.6) has a unique solution, i.e., we have obtained a contradiction.

Let us prove the converse. Let $\bar{x}$ be an extreme point of the polyhedron M. Evidently,

$$J(\bar{x}) = \{j \mid (a_j, \bar{x}) = b_j\} \ne \varnothing.$$

Consider the k-boundary system

$$\begin{cases} (a_i, x) = b_i, & \forall i \in \bar{J} \subset J(\bar{x}), \\ (a_j, x) \le b_j, & \forall j \notin \bar{J}, \end{cases} \tag{2.7}$$

i.e., $|\bar{J}| = k$, $\{a_i\}_{i \in \bar{J}}$ are linearly independent. Evidently, the vector $\bar{x}$ solves this system. The fact is that $k = n$. Indeed, if $k < n$, then we can choose two different points p and q in the linear manifold $(a_i, x) = b_i$, $i \in J(\bar{x})$, such that $\bar{x} = (p + q)/2$, $(a_j, p) < b_j$, $(a_j, q) < b_j$, $\forall j \notin J(\bar{x})$, which contradicts the definition of an extreme point. So we have established the equivalence of the two notions: a vertex and an extreme point.

3. BOUNDED POLYHEDRONS

Assume that a system of linear inequalities

$$(a_j, x) \le b_j, \qquad j = 1, \ldots, m, \tag{3.1}$$

or, in the matrix form,

$$Ax \le b,$$

sets a polyhedron $M \subset \mathbb{R}^n$. Below we shall give several equivalent conditions of boundedness of M. By the *boundedness* of M we mean fulfilment of the condition $\sup_{x \in M} \|x\| < +\infty$, where $\|\cdot\|$ is the Euclidean norm. This condition is equivalent (as will be shown below) to the following one: for arbitrary $p \in M$ and $s \neq 0$ the ray $\{p + ts\}_{t \geq 0}$ does leave M. This means that for some $\bar{t}(s)$: $p + \bar{t}(s)s \notin M$. This condition may be laid into foundation of the definition of the boundedness of a polyhedron M without introducing the norm if, instead of $\mathbb{R}^n$, we consider the n-dimensional arithmetic space $\mathbb{E}_n$. This has certain sense because we consider the *algebraic* theory of polyhedrons, and within the limits of this theory it is not necessary to introduce one or another norm in $\mathbb{E}_n$.

Let us introduce some definitions.

Definition 3.1. A system of vectors $\{a_j\}_1^m$ is called *multifold* if for each $s \in \mathbb{E}_n$, $s \neq 0$, $\exists j'$: $(a_{j'}, s) > 0$.

Definition 3.2. A system of vectors $\{a_j\}_1^m$ is called a *nonnegative basis* of the space $\mathbb{E}_n$ if $\mathrm{cone}\{a_j\}_1^m = \mathbb{E}_n$. Here $\mathrm{cone}\{a_j\}_1^m$ is the conic hull of the system $\{a_j\}_1^m$, i.e., the set of all nonnegative linear combinations of these vectors.

Definition 3.3. A direction $s \neq 0$ is called *recessive* for M if for some $p \in M$ the ray $\{p + ts\}_{t \geq 0}$ belongs to M.

The set of all recessive directions (with zero) forms a cone and coincides with the set of solutions of the homogeneous system of linear inequalities

$$(a_j, x) \leq 0, \qquad j = 1, \ldots, m, \tag{3.2}$$

that corresponds to system (3.1). This cone is conventionally denoted as $\mathrm{rec}\, M$.

Let us check the above-mentioned property. If $s \in \mathrm{rec}\, M$, then for $p \in M$ we have $A(p + ts) \leq b$, $\forall t \geq 0$, i.e., $Ap - b + tAs \leq 0$, $\forall t \geq 0$. Hence it follows that $As \leq 0$. Conversely, if $s \neq 0$ and $As \leq 0$, then for $p \in M$ and for each $t \geq 0$ we have $A(p + ts) \leq b$, i.e., $s \in \mathrm{rec}\, M$.

Now we consider three properties that system (3.1) may have:

1. $\mathrm{cone}\{a_j\}_1^m = \mathbb{E}_n$;

2. $\{a_j\}_1^m$ is a multifold system;

3. $\mathrm{rec}\, M = \{0\}$.

Theorem 3.1. *Properties 1–3 are equivalent.*

Proof. We shall follow the scheme $3 \to 2 \to 1 \to 3$.

$3 \to 2$. If 2 does not hold, then $\exists s \neq 0 : As \leq 0$, which contradicts property 3.

$2 \to 1$. Suppose that 1 does not hold, i.e., $K := \text{cone}\{a_j\}_1^m \neq \mathbb{E}_n$. We can construct a hyperplane of support to the convex cone K, the equation for the hyperplane being $(s, x) = 0$, $s \neq 0$, such that $x \in K \Rightarrow (s, x) \leq 0$. Hence it follows that $(s, a_j) \leq 0$ (j), i.e., $As \leq 0$, which contradicts property 2.

$1 \to 3$. Suppose that 3 does not hold. Then $\exists s \neq 0 : As \leq 0$. By property 1, we have $s = \sum_{j=1}^m \lambda_j a_j^T = A^T \lambda$, $\lambda \geq 0$. Therefore $AA^T \lambda \leq 0$. Multiplying this inequality by $\lambda \geq 0$ we obtain

$$(\lambda, AA^T \lambda) = (A^T \lambda, A^T \lambda) = (s, s) = \|s\| \leq 0,$$

i.e., $s = 0$. Again we have got a contradiction. $\quad \square$

Theorem 3.2. *The polyhedron M of system* (3.1) *is bounded if and only if* $\text{rec } M = \{0\}$.

Proof. *Necessity.* If M is bounded, then it contains no rays $\{p + ts\}_{t \geq 0}$, $s \neq 0$, which implies $\text{rec } M = \{0\}$.

Sufficiency. If $M \in \mathbb{R}^n$ is unbounded, then there exists a sequence of points $\{z_k = p + t_k s_k\} \subset M$ such that $p \in M$, $\|s_k\| = 1$, $\forall k$, $t_k \to +\infty$ $(t_k = \|z_k - p\|)$. From the sequence $\{s_k\}$, a converging subsequence can be chosen. Without loss of generality, let it be $\{s_k\}$ itself, $\{s_k\} \to s$. Since $Az_k \leq b$ or, in other words, $t_k As_k \leq b - Ap$, therefore, $As \leq 0$ because $t_k \to +\infty$. But in this case the ray $\{p + ts\}_{t \geq 0}$ will belong to the polyhedron M: $A(p + ts) = Ap + tAs \leq b$. This means that s is a recessive direction, a contradiction. $\square$

Remark 3.1. Since condition 3 is equivalent to each of the two conditions 1 and 2, therefore, Theorem 3.2 can be formulated with any of these two conditions. Moreover, there exist other variants equivalent to conditions considered in Theorem 3.1, for example,

1°) $r(A) = n$ and $\exists \bar{u} > 0$: $A^T \bar{u} = 0$;

2°) $\forall c \neq 0 \ \exists \tilde{u} > 0$: $c = A^T \tilde{u}$.

Really, if $\text{cone}\{a_j\} = \mathbb{E}_n$ (in this case $r(A) = n$), then from $-\sum_{j=1}^{m} a_j = \sum_{j=1}^{m} \lambda_j a_j$, $\lambda_j \geq 0$ ($\forall j$), it follows that $A^T \bar{u} = \sum_{j=1}^{m} \bar{u}_j a_j = 0$ for $\bar{u}_j = 1 + \lambda_j > 0$; i.e., the implication $(\text{cone } A = \mathbb{E}_n) \Rightarrow 1^\circ$ is true.

Further, for each $c \neq 0$ we have $c = A^T \bar{v}$ for a certain $\bar{v} \in \mathbb{E}_m$. Hence $c = A^T (\bar{v} + \alpha \bar{u})$, which, for $\alpha > 0$ sufficiently large, gives $\bar{v} + \alpha \bar{u} =: \tilde{u} > 0$. Thus, condition 1° implies condition 2°. Finally, condition 2° immediately implies $\text{cone}\{a_j\} = \mathbb{E}_n$; therefore, each of the conditions 1°, 2° is equivalent to each of the conditions 1–3 from Theorem 3.1.

In connection with condition 1, we consider now the question about a minimal nonnegative basis, i.e., a basis with the minimal number of vectors. Since a nonnegative basis is evidently a simple basis, therefore, the minimal basis contains at least n vectors. But it cannot contain n vectors. To obtain a contradiction, assume that there exists a nonnegative basis of n vectors: $a_1, a_2, \ldots, a_n$. In this case $\{a_j\}_1^n$ are linearly independent. Take the vector $a = -\sum_{j=1}^{n} a_j$ and write its nonnegative representation in terms of the basis $\{a_j\}_1^n$

$$-\sum_{j=1}^{n} a_j = \sum_{j=1}^{n} \alpha_j a_j, \quad a_j > 0, \quad \forall j, \quad \text{i.e.,} \quad \sum_{j=1}^{n} (\alpha_j + 1) a_j = 0.$$

This is a contradiction since $\alpha_j + 1 > 0$ whereas $\{a_j\}_1^n$ are linearly independent.

In reality, each minimal nonnegative basis contains $(n+1)$ vectors. Construction of such a basis is trivial. Let us do it.

If $\{e_j\}_1^n$ is a usual basis in $\mathbb{E}_n$, then the collection of vectors $\{e_j\}_1^{n+1}$ with $e_{n+1} = -\sum_{j=1}^{n} e_j$ is a nonnegative basis. Indeed, let a be an arbitrary vector from $\mathbb{R}^n$, then $a = \sum_{j=1}^{n} \alpha_j e_j$. Taking into account the identity $\alpha e_{n+1} + \sum_{j=1}^{n} \alpha e_j = 0$ valid for all α, we may write $a = \alpha e_{n+1} + \sum_{j=1}^{n} (\alpha_j + \alpha) e_j$. Taking $\alpha > \max_{j \in \overline{1,n}} |\alpha_j|$, we obtain a nonnegative decomposition of the vector a in terms of $\{e_j\}_1^{n+1}$.

If $\{a_j\}_1^{n+1}$ is a minimal nonnegative basis, then any n its vectors are linearly independent. For definiteness, let the subsystem $\{a_j\}_1^n$ be linearly independent. We need to show that any other subsystem of n vectors, say $\{a_j\}_2^{n+1}$, is linearly independent. Assume that the latter system is linearly dependent, then $a_{n+1} = \sum_{i=2}^{n} \gamma_i a_i$. Consider the decomposition

$$-a_1 = \sum_{j=1}^{n+1} \alpha_j a_j, \qquad \alpha_j \geq 0, \qquad j = 1, \ldots, n+1,$$

which together with the previous equality gives

$$(\alpha_1 + 1)a_1 + \sum_{j=2}^{n}(\alpha_j + a_{n+1}\gamma_j)a_j = 0.$$

Since the vectors $\{a_j\}_1^n$ are linearly independent, therefore, $\alpha_1 + 1 = 0$, which contradicts the condition $\alpha_1 \geq 0$.

The above considerations are summarized in a theorem below.

Theorem 3.3. *A minimal nonnegative basis of the space $\mathbb{E}_n$ consists of exactly $(n + 1)$ vectors, and any n of them are linearly independent.*

Further we shall need the following theorem.

Theorem 3.4. *If $a \in \text{cone}\{a_j\}_1^m$, then $a \in \text{cone}\{a_j\}_{j \in J}$ with linearly independent $\{a_j\}$, $j \in J$.*

Proof. From all representations of the vector a in the form of a nonnegative linear combination of the vectors $\{a_j\}$ we choose that representation which has the minimal number of positive coefficients. For definiteness, let it be

$$a = \sum_{j=1}^{m} \lambda_j a_j. \tag{3.3}$$

Let us show that the vectors $\{a_j\}_1^m$ are linearly independent. Assuming the contrary, we have

$$\sum_{j=1}^{m} \alpha_j a_j = 0, \tag{3.4}$$

where not all of α_j's are zeros. Suppose, for definiteness, that $\alpha_1 \neq 0$, moreover, let $\alpha_1 < 0$. Consider the combination $(3.3) + t(3.4)$, which is an identity with respect to the parameter t:

$$a = \sum_{j=1}^{m} (\lambda_j + t\alpha_j)a_j. \tag{3.5}$$

Suppose that t^* is such that

$$\max\{t \mid \lambda_j + t\alpha_j \geq 0, \ j = 1, \ldots, m\} = \min_{\alpha_j < 0} \frac{\lambda_j}{|\alpha_j|} = \frac{\lambda_{j_*}}{|\alpha_{j_*}|} =: t^*.$$

If we substitute this t^* into (3.5), then all the coefficients of the representation will be nonnegative, and the coefficient with number j_* will be equal to zero. So, we have obtained a nonnegative expansion of the vector a in terms of $\{a_j\}$ in which the number of positive coefficients is less than m. This contradicts the assumption that representation (3.3) is minimal. $\square$

Corollary 3.1. *An arbitrary vector a from the convex hull $\mathrm{co}\{a_j\}_1^m$ can be represented in the form*

$$a = \sum_{j \in J} \alpha_j a_j, \qquad \alpha_j \geq 0, \qquad \sum_{j \in J} \alpha_j = 1, \qquad |J| \leq n+1.$$

Proof. Really, the inclusion $a \in \mathrm{co}\{a_j\}_1^m$ means that

$$a = \sum_{j=1}^m \beta_j a_j, \qquad \beta_j \geq 0, \quad j = 1, \ldots, m, \qquad \sum_{j=1}^m \beta_j = 1.$$

Let

$$\bar{a} = \begin{bmatrix} a \\ 1 \end{bmatrix}, \qquad \bar{a}_j = \begin{bmatrix} a_j \\ 1 \end{bmatrix},$$

then the last relations yield

$$\bar{a} = \sum_{j=1}^m \beta_j \bar{a}_j \in \mathrm{cone}\{\bar{a}_j\}_1^m.$$

By Theorem 3.4, there exists a representation for $\bar{a}$ of the form

$$\bar{a} = \sum_{j=1}^m \alpha_j \bar{a}_j, \qquad \alpha_j \geq 0, \qquad \sum_{j=1}^m \alpha_j = 1,$$

where $\{\bar{a}_j \mid \alpha_j > 0\}$ are linearly independent. Since the rank of the matrix

$$\begin{bmatrix} a_1 & a_2 & \cdots & a_m \\ 1 & 1 & \cdots & 1 \end{bmatrix}$$

is no greater than $n+1$, therefore, there are no more than $n+1$ terms with $\alpha_j > 0$ in the equality obtained above. This means that

$$\bar{a} = \sum_{j \in J} \alpha_j \bar{a}_j, \qquad |J| \leq n+1,$$

or

$$a = \sum_{j \in J} \alpha_j a_j, \qquad \alpha_j \geq 0 \quad (j), \qquad \sum_{j \in J} \alpha_j = 1.$$

□

4. A PARAMETRIC REPRESENTATION OF POLYHEDRONS

For compatible systems of linear equations, the following result is well-known from linear algebra: *an arbitrary solution of such a system can be represented as the sum of a particular solution and an arbitrary linear combination of some linearly independent collection of solutions of the corresponding homogeneous system.* There is an analog of this result for systems of linear inequalities, namely: *an arbitrary solution $\bar{x}$ of a system of linear inequalities*

$$(a_j, x) \leq b_j, \qquad j = 1, \ldots, m, \tag{4.1}$$

of rank $r = n$ is representable in the form

$$\bar{x} = \sum_{j \in J} \alpha_j p_j + \sum_{i \in I} \lambda_i s_i. \tag{4.2}$$

Here $\{p_j\}_J$ is the collection of the vertices of the polyhedron M of system (4.1), $\alpha_j \geq 0$, $j = 1, \ldots, m$, $\sum_J \alpha_j = 1$; $\{s_i\}_I$ is the set of directrices of unbounded edges (rays) of the polyhedron M, $\lambda_i \geq 0$, $i = 1, \ldots, m$. Relation (4.2) can be written as

$$M = \mathrm{co}\{p_j\} + \mathrm{cone}\{s_i\}.$$

The present section is devoted to consideration of the above result, moreover, we shall also treat the case when $r < n$.

First, let us consider the case when M is a bounded polyhedron; i.e., the system of vectors $\{a_j\}_1^m$ is multifold.

Lemma 4.1. *Let the polyhedron M be bounded and $\{p_j\}_J$ be the set of its vertices. Then*

$$M = \mathrm{co}\{p_j\}_J \quad (=: N). \tag{4.3}$$

Proof. The inclusion $N \subset M$ is trivial. Let us prove the reverse inclusion. Suppose that $p \in M$. The point p belongs to a certain set of k-faces M_k of the polyhedron M. We shall prove the inclusion $p \in N$ by induction on k decrementing it from n to 0, i.e., $k = n, \ldots, 1, 0$, in the sense of the property of belonging $p \in M_k$. If $k = n$, then $p \in M_n$. But an n-face is a vertex; therefore, $p \in N$. Here the 0-face is understood as the polyhedron M itself, i.e., $M_0 = M$.

Assume that we have proved the necessary inclusion for $s = n, \ldots, k$ ($k > 0$), i.e., $p \in M_k \Rightarrow p \in N$. Let us prove it for the case $p \in M_{k-1}$. To this end, we choose a point q_1 (using the rule from the proof of Theorem 2.1) such that the ray $\{p + t(q_1 - p)\}_{t \geq 0}$ leaves M at the point p' which belongs to a certain k-face $M_k' \subset M$. For the opposite ray, the conditions of choice of the first ray will hold true (see the proof of Theorem 2.1). Therefore, the point p'' at which the ray $\{p + t(p - q_1)\}_{t \geq 0}$ leaves M will also belong to some k-face M_k''. So, we have obtained the following situation $p \in \overline{p', p''}$, $p' \in M_k'$, $p'' \in M_k''$. By the induction assumption, $p' \in N$, $p'' \in N$. But N is a convex set, therefore, $p \in N$. Since $p \in M$ was arbitrary, therefore, $M \subset N$. $\square$

The homogeneous system corresponding to system (4.1) is the system

$$(a_j, x) \leq 0, \quad j = 1, \ldots, m; \tag{4.1_0}$$

it defines a convex cone K. We assume that $r = n$ and $K \neq \{0\}$, i.e., the system of vectors $\{a_j\}_1^m$ is not multifold.

Lemma 4.2. *The equality*

$$K = \text{cone}\{s_i\}_I \quad (=: S) \tag{4.4}$$

holds, where s_i are the directrices of the edges of the cone K (i.e., $s_i \neq 0$ and s_i belong to $(n-1)$-faces of the cone K).

Proof. The construction of $\{s_i\}_I$ is be as follows. Consider the system

$$(a_j, x) \leq 0, \quad j = 1, \ldots, m, \quad -\left(\sum_{j=1}^m a_j, x\right) \leq 1. \tag{4.5}$$

The set of vectors $\{a_1, \ldots, a_m, a_{m+1} = -\sum_{j=1}^m a_j\}$ is multifold (see Section 3); therefore, the polyhedron, say K', of system (4.5) is bounded. Let $\{s_i\}_I$ be the set of all its vertices except 0. By construction, $S \subset K$. Now we need to show the reverse inclusion $K \subset S$.

As a preliminary, let us make several remarks concerning the polyhedron K' of system (4.5). Each its vertex s_i is a unique solution of a certain system

$$\begin{cases} (a_{j_t}, x) = 0, \quad t = 1, \dots, n-1, \quad (a_{m+1}, x) = 1, \\ (a_j, x) \le 0, \quad j \ne j_1, \dots, j_{n-1}, \end{cases} \tag{4.6}$$

where $\{a_{j_t}, t = 1, \dots, n-1, a_{m+1}\}$ are linearly independent (see the definition of a k-face, Section 2). The index $m+1$ necessarily appears in the n-boundary system (4.6), for otherwise s_i would be a solution of a homogeneous system of rank n, which would imply $s_i = 0$. Since s_i satisfies the first $n-1$ equations of system (4.6), therefore, s_i is the direction vector of the $(n-1)$-face (edge) defined by these equations. Note also that all the vertices of the polyhedron K' except zero are the vertices of the polyhedron K'' of the system

$$(a_j, x) \le 0, \quad j = 1, \dots, m, \quad (a_{m+1}, x) = 1. \tag{4.7}$$

By Lemma 4.1, we have $\mathrm{co}\{s_i\}_I = K''$.

Now we proceed to proving that $K \subset S$. Suppose that $p \in K$ and $p \ne 0$. Consider the ray $\{pt\}_{t \ge 0} \subset K$. It intersects the hyperplane $(a_{m+1}, x) = 1$ at the point $\bar{x} = \bar{t}p$, $\bar{t} = \left(\sum_{j=1}^m |(a_j, p)| \right)^{-1}$. Here $\sum_{j=1}^m |(a_j, p)| \ne 0$, for otherwise we would have $p = 0$. Since $\bar{x} \in K'' = \mathrm{co}\{s_i\}_I$, it follows that $p = \bar{x}/\bar{t} \in \mathrm{cone}\{s_i\}_I$. Therefore, $K \subset S$. Thus, equality (4.4) is proved. $\square$

Theorem 4.1. *Let system (4.1) be compatible and its rank be $r = n$. Let $\{p_j\}_J$ be all the vertices of the polyhedron M of system (4.1) and $\{s_i\}_I$ be the directrices of the edges (i. e., of the $(n-1)$-faces) of the cone of solutions of the corresponding system of inequalities (4.1_0). Then*

$$M = \mathrm{co}\{p_j\}_J + \mathrm{cone}\{s_i\}_I \quad (=: N). \tag{4.8}$$

Proof. The inclusion $N \subset M$ is trivial. The reverse inclusion is proved analogously to Lemma 4.1. Namely, let $p \in M$. If $p \in M_n$, then p is a vertex of the polyhedron M; therefore, $p \in N$. Suppose that for $s = n, \dots, k$ we have already proved the implication $p \in M_s \Rightarrow p \in N$, where M_s is some s-face of the polyhedron M. Let us prove this implication for $p \in M_{k-1}$. We choose q_1 following the rule from the proof of Theorem 2.1, i. e., we choose q_1 such that the ray $\{p + t(q_1 - p)\}_{t \ge 0}$ leaves M at a point p' which belonging to some k-face $M_k' \subset M$. As concerns the opposite ray $\{p + t(p - q_1)\}_{t \ge 0} =: \Gamma$, two cases are possible. If it leaves M, then, as in the proof of Lemma 4.1, the leaving point q'' will belong to some k-face $M_k'' \subset M$. Then the conclusion that $p \in N$ is drawn completely by analogy with the above-mentioned proof. If $\Gamma \subset M$, then the direction $s = p - q_1$ is recessive, i.e., $s \in K = \mathrm{cone}\{s_i\}_I$, $s \neq 0$. The point p lies on the ray $\{q_1 + ts\}_{t \geq 0}$, i.e., $p = q_1 + \bar{t}s$ ($\bar{t} = 1$). Since $q_1 \in M_k' \subset N$ (by the induction assumption) and $s \in \mathrm{cone}\{s_i\}_I$, therefore,

$$p \in \mathrm{co}\{p_j\}_J + \mathrm{cone}\{s_i\}_I.$$

Hence $p \in N$, which completes the proof of the theorem. $\square$

Remark 4.1. We have singled out the case $r = n$ and presented it as a theorem because in this case the vectors $\{p_j\}$ and $\{s_i\}$ in expansion (4.8) admit of a clear geometric interpretation. If $r < n$, then the third term appears in formula (4.8) and the sets of vectors $\{p_j\}$ and $\{s_i\}$ are interpreted in a more complicated manner.

Now we consider system (4.1) in the general case. Let H denote the set of solutions of the homogeneous system (the kernel)

$$(a_j, x) = 0, \qquad j = 1, \ldots, m. \tag{4.9}$$

This set is a subspace of the space $\mathbb{E}_n$ of dimension $n - r$. The space $\mathbb{E}_n$ can be represented as the direct sum

$$\mathbb{E}_n = \mathbb{E}_r + H, \tag{4.10}$$

where $\mathbb{E}_r$ is a subspace of dimension r. Similar to Section 1, having taken a base $\{e_1, \ldots, e_r\}$ of the subspace $\mathbb{E}_r$, we can write any vector $x \in \mathbb{E}_n$ in the form

$$x = \sum_{i=1}^{r} y_i e_i + h, \qquad h \in H.$$

Substituting x into (4.1) we obtain

$$\sum_{i=1}^{r} (a_j, e_i) y_i \leq b_j, \qquad j = 1, \ldots, m,$$

or

$$(\bar{a}_j, y) \leq b_j, \qquad j = 1, \ldots, m, \tag{4.11}$$

where $\bar{a}_j^T = [a_{j1}, \ldots, a_{jr}]$, $a_{ji} = (a_j, e_i)$.

The rank of system (4.11) is r, which coincides with the dimension of $\mathbb{E}_r$, the latter being the space relative to which system (4.11) is written.

Let $\{\bar{p}_j\}_J$ and $\{\bar{s}_i\}_I$ be the set of vertices of the polyhedron M^r of system (4.11) and the set of directrices of edges of the homogeneous system $(\bar{a}_j, y) \leq 0$, $j = 1, \ldots, m$, respectively. By Theorem 4.1, we have

$$M^r = \mathrm{co}\{\bar{p}_j\}_J + \mathrm{cone}\{\bar{s}_i\}_I. \tag{4.12}$$

Let the coordinates of $\bar{p}_j$ and $\bar{s}_i$ be as follows:

$$\bar{p}_j = [p_j^1, \ldots, p_j^r]^T, \qquad \bar{s}_i = [s_i^1, \ldots, s_i^r]^T.$$

If we set

$$p_j = \sum_{i=1}^r p_j^i e_i, \qquad s_i = \sum_{k=1}^r s_i^k e_k,$$

then the set represented by the algebraic sum $\mathrm{co}\{p_j\}_J + \mathrm{cone}\{s_i\}_I$ will be none other than the projection of the polyhedron M of system (4.1) on the subspace $\mathbb{E}_r \subset \mathbb{E}_n$ and for the polyhedron M itself the following equality will hold:

$$M = \mathrm{co}\{p_j\}_J + \mathrm{cone}\{s_i\}_I + H. \tag{4.13}$$

Thus, we have proved

Theorem 4.2. *For the polyhedron M of a compatible system of inequalities (4.1) there exist collections of vectors $\{p_j\}_J \subset \mathbb{E}_n$ and $\{s_i\}_I \subset \mathbb{E}_n$ such that*

$$M = \mathrm{co}\{p_j\}_J + \mathrm{cone}\{s_i\}_I + H,$$

where H is the kernel of the system of functionals $\{(a_j, x)\}_1^m$.

Remark 4.2. With such a formulation of the polyhedron decomposition theorem, an arbitrary real linear space $\mathbb{X}$ may be taken as the initial space. The formulation of the theorem and its proof will be just the same.

5. THE FARKAS – MINKOWSKI THEOREM ON DEPENDENT INEQUALITIES

Theorem 5.2 to be proved below is the central theorem in Chapter 1.

Definition 5.1. We shall say that a linear inequality

$$(c, x) - \alpha \leq 0 \tag{5.1}$$

is *dependent* on a compatible system of linear inequalities

$$(a_j, x) - b_j \leq 0, \qquad j = 1, \ldots, m, \tag{5.2}$$

(or is its *implication*) if each solution of system (5.2) satisfies inequality (5.1).

Let us consider the condition of dependence of (5.1) on (5.2) for the case $\alpha = b_j = 0$, $j = 1, \ldots, m$.

Theorem 5.1 [Farkas – Minkowski, homogeneous case]. *An inequality*

$$(c, x) \leq 0 \tag{5.1_0}$$

is an implication of a system

$$(a_j, x) \leq 0, \qquad j = 1, \ldots, m, \tag{5.2_0}$$

if and only if there exist $\lambda_j \geq 0$, $j = 1, \ldots, m$, *such that*

$$c = \sum_{j=1}^{m} \lambda_j a_j, \tag{5.3}$$

i.e., $(a, x) \equiv \sum_{j=1}^{m} \lambda_j (a_j, x)$.

Proof. The sufficiency is obvious. The necessity will be proved by induction on the number m.

Let $m = 1$. The inequality $(c, x) \leq 0$ is obviously an implication of the equation $(a_1, x) = 0$. Then, by Lemma 2.1, we have $(a, x) \equiv \lambda(a_1, x)$ for some $\lambda \in \mathbb{R}$. If we take a point p such that $(a_1, p) < 0$, then the inequality $(a, p) \leq 0$ should hold. Hence it follows that $\lambda \geq 0$. Consequently, the necessity is proved for $m = 1$.

We suppose now that the necessity has been proved for systems with the number of inequalities less than m. Write the system

$$(a_j, x) = 0, \qquad j = 1, \ldots, m. \tag{5.4}$$

Evidently, inequality (5.1_0) is its implication; therefore, by Lemma 2.1, we have

$$(c, x) \equiv \sum_{j=1}^{m} \lambda_j (a_j, x) \tag{5.5}$$

for some $\lambda_j \in \mathbb{R}$, $j = 1, \ldots, m$. We need to prove that there exists a representation (5.5) such that $\lambda_j \geq 0$, $j = 1, \ldots, m$. If none such representation exists, then among all the representations (5.5) we choose one of those for which the number of negative coefficients is minimal. Suppose that (5.5) is such a representation. Without any loss of generality, we assume that the negative coefficients are the first s coefficients $\lambda_1, \ldots, \lambda_s$.

Now consider the linear functional

$$(e, x) \equiv \lambda_1 (a_1, x) + \sum_{j=s+1}^{m} \lambda_j (a_j, x). \tag{5.6}$$

Taking into account (5.5) and (5.6) we have

$$(e, x) \equiv (c, x) - \sum_{j=2}^{s} \lambda_j (a_j, x). \tag{5.7}$$

Since $\lambda_j < 0$, $j = 1, \ldots, s$, therefore, (5.7) and the dependence of inequality (5.1_0) on system (5.2_0) imply that the inequality

$$(e, x) \leq 0 \tag{5.8}$$

is an implication of the system

$$(a_1, x) \leq 0, \qquad (a_j, x) \leq 0, \quad j = 2, \ldots, m.$$

It follows from (5.6) that (5.8) is an implication of the system

$$-(a_1, x) \leq 0, \qquad (a_j, x) \leq 0, \quad j = 2, \ldots, m.$$

Therefore, the inequality $(e, x) \leq 0$ is an implication of the system

$$(a_j, x) \leq 0, \qquad j = 2, \ldots, m,$$

but then, by the induction assumption, we have

$$(e, x) \equiv \sum_{j=2}^{m} \mu_j (a_j, x)$$

for some $\mu_j \geq 0$, $j = 2, \ldots, m$. The last relation and (5.7) yield

$$(c, x) \equiv \sum_{j=2}^{s} (\lambda_j + \mu_j)(a_j, x) + \sum_{j=s+1}^{m} \mu_j(a_j, x). \tag{5.9}$$

In this representation, the number of negative coefficients before (a_j, x) is less than s, which contradicts the choice of s. Therefore, the assumption that there is none representation (5.5) with $\lambda_j \geq 0$, $j = 1, \ldots, m$, has led to a contradiction. The theorem is proved.　□

Before we formulate the Farkas – Minkowski theorem for the general case, i. e., for the case (5.1), (5.2), we shall prove the following lemma:

Lemma 5.1. *Inequality* (5.1) *is an implication of a compatible system* (5.2) *if and only if the inequality*

$$(c, x) - \alpha t \leq 0 \tag{5.10}$$

is an implication of the system

$$\begin{cases} (a_j, x) - b_j t \leq 0, & j = 1, \ldots, m, \\ -t \leq 0. \end{cases} \tag{5.11}$$

Proof. Note that inequalities (5.10) and (5.11) are homogeneous relative to the variable $[x, t] \in \mathbb{X} \times \mathbb{R}^1$. The sufficiency of the lemma conditions is evident. Now we need to show that if (5.1) is an implication of (5.2), then (5.10) is an implication of (5.11). Let $[x_0, t_0]$ solve system (5.11). If $t_0 > 0$, then the vector x_0/t_0 solves system (5.2); therefore, by the assumption, it solves inequality (5.1). Hence $[x_0, t_0]$ satisfies inequality (5.10). It remains to consider the case $t_0 = 0$. Let $[\bar{x}, \bar{t}]$ be some solution of system (5.11) with $\bar{t} > 0$. For example, $[\bar{x}, 1]$ where $\bar{x}$ is an arbitrary solution of system (5.2) is such a solution. For each $0 < \gamma < 1$, the vector

$$z(\gamma) = (1 - \gamma)[\bar{x}, \bar{t}] + \gamma[x_0, t_0] = [(1 - \gamma)\bar{x} + \gamma x_0, (1 - \gamma)\bar{t} + \gamma t_0]$$

solves system (5.11). But since $(1 - \gamma)\bar{t} + \gamma t_0 > 0$, therefore, in view of the case considered above, we obtain

$$(c, x(\gamma)) - \alpha t(\gamma) \leq 0,$$

where $x(\gamma) = (1 - \gamma)\bar{x} + \gamma x_0$, $t(\gamma) = (1 - \gamma)\bar{t} + \gamma t_0$. We have

$$(c, x(\gamma)) - \alpha t(\gamma) = (1 - \gamma)[(c, \bar{x}) - \alpha\bar{t}] + \gamma[(c, x_0) - \alpha t_0] \leq 0.$$

Passing to the limit as $\gamma \to 1$ in the above inequality, we obtain

$$(c, x_0) - \alpha t_0 \leq 0,$$

i.e., the solution $[x_0, t_0]$ of system (5.11) for $t_0 = 0$ satisfies inequality (5.10).
□

Theorem 5.2 [Farkas – Minkowski]. *Inequality* (5.1)

$$(c, x) - \alpha \leq 0$$

is an implication of a compatible system (5.2)

$$(a_j, x) - b_j \leq 0, \qquad j = 1, \ldots, m,$$

if and only if for some $\lambda_j \leq 0$, $j = 0, 1, \ldots, m$, the following relation takes place:

$$(c, x) - \alpha \equiv \sum_{j=1}^{m} \lambda_j [(a_j, x) - b_j] - \lambda_0. \tag{5.12}$$

Proof. The sufficiency of condition (5.12) is evident. Suppose now that inequality (5.1) is an implication of system (5.2). Then, by Lemma 5.1, inequality (5.10) is an implication of system (5.11). By Theorem 5.1, there exist $\lambda_j \geq 0$, $j = 0, 1, \ldots, m$, such that

$$(c, x) - \alpha t \equiv \sum_{j=1}^{m} \lambda_j [(a_j, x) - b_j t] - \lambda_0 t. \tag{5.13}$$

This relation for $t = 1$ yields (5.12). □

We shall classify the dependent inequalities following the next definition.

Definition 5.2. Inequality (5.1) dependent on (5.2) shall be called an *implication of first kind* if for some $\varepsilon > 0$ the inequality $(c, x) - \alpha \leq -\varepsilon$ is also dependent on system (5.2). Otherwise, we shall call inequality (5.1) an *implication of second kind*.

Remark 5.1. If inequality (5.1) is a first-kind implication of system (5.2), then we may assume that $\lambda_0 > 0$ in (5.12). Really, since for some $\varepsilon > 0$ the inequality

$$(c, x) - \alpha + \varepsilon \leq 0$$

is an implication of system (5.2), therefore, for some $\lambda_j \geq 0$, $j = 1, \ldots, m$, $\bar{\lambda} \geq 0$, the relation

$$(c, x) - \alpha + \varepsilon \equiv \sum_{j=1}^{m} \lambda_j [(a_j, x) - b_j] - \bar{\lambda}$$

holds true. If we set $\lambda_0 = \bar{\lambda} + \varepsilon$, then the above relation gives the required relation.

If inequality (5.1) is a second-kind implication of system (5.2), then $\lambda_0 = 0$ in (5.12). Really, if $\lambda_0 > 0$, then, by Theorem 5.2, the inequality $(c, x) - \alpha + \lambda_0 \leq 0$ is an implication of system (5.2), which contradicts the definition of the dependence of second kind.

The next result follows immediately from the definition of the second-kind implication.

Theorem 5.3. *Inequality* (5.1) *is a second-kind implication of system* (5.2) *if and only if*

$$\alpha = \sup\{(c, x) \mid (a_j, x) \leq b_j, \ j = 1, \ldots, m\}.$$

6. ATTAINABILITY THEOREM FOR INEQUALITIES-IMPLICATIONS OF SECOND KIND

Consider a compatible system

$$Ax \leq b \tag{6.1}$$

and an inequality

$$(c, x) \leq \alpha. \tag{6.2}$$

Lemma 6.1. *System* (6.1) *is incompatible if and only if the inequality* $t \leq 0$ *is an implication of the system*

$$Ax \leq bt. \tag{6.1_t}$$

Proof. *Necessity.* We suppose that system (6.1) is incompatible. If in this case the inequality $t \leq 0$ is not an implication of system (6.1_t), then the system has a solution $[\bar{x}, \bar{t}]$, $\bar{t} > 0$. Hence it follows that $\bar{x}/\bar{t}$ solves system (6.1), which contradicts the assumption of its incompatibility.

Sufficiency. Suppose that the inequality $t \leq 0$ is an implication of system (6.1_t). If in this case system (6.1) is compatible and $\bar{x}$ is some of its solutions, then the vector $[\bar{x}, 1]$ solves system (6.1_t). However, the latter solution does not satisfy the inequality $t \leq 0$, which contradicts the assumption. $\square$

Theorem 6.1. *If inequality (6.2) is a second-kind implication of system (6.1), then the system*

$$\begin{cases} (c, x) = \alpha, \\ \quad Ax \leq b \end{cases} \tag{6.3}$$

is compatible.

Proof. Assuming the contrary, we obtain the system

$$\begin{cases} -(c, x) + \alpha \leq 0, \\ \quad Ax \leq b \end{cases} \tag{6.4}$$

that is also incompatible. By Lemma 6.1, the inequality $t \leq 0$ is an implication of the system

$$\begin{cases} -(c, x) + \alpha t \leq 0, \\ \quad Ax \leq bt. \end{cases} \tag{6.4_t}$$

By the Farkas – Minkowski theorem, we have

$$t \stackrel{(x,t)}{\equiv} (Ax - bt, \lambda) + \lambda_0[-(c, x) + \alpha t] \tag{6.5}$$

for some $\lambda \geq 0$, $\lambda_0 \geq 0$. In reality, $\lambda_0 > 0$, for otherwise the inequality $t \leq 0$ would be an implication of system (6.1_t), which, by Lemma 6.1, would imply that system (6.1) is incompatible.

If we set now $t = 1$ in relation (6.5), then we can rewrite it as

$$(c, x) - \alpha + \frac{1}{\lambda_0} \stackrel{(x)}{\equiv} \left(Ax - b, \frac{\lambda}{\lambda_0}\right).$$

Therefore, the inequality $(c, x) - \alpha \leq -\varepsilon = -\lambda_0^{-1}$ is an implication of system (6.1), i.e., (6.2) is a first-kind implication of system (6.1), which contradicts the theorem assumption. $\square$

It was noted in Section 5 that α from (6.2) in the situation in question is the optimal value of the problem of linear programming

$$L : \sup\{(c, x) \mid Ax \le b\}. \tag{6.6}$$

In this connection we can reformulate Theorem 6.1 as follows:

If in the problem L the function (c, x) is bounded from above on $M = \{x \mid Ax \le b\}$, then the operation sup *in it is attainable, i. e.,*

$$\exists \bar{x} \in M : (c, \bar{x}) = \alpha = \max_{x \in M}(c, x).$$

The set of such $\bar{x}$'s is called the *optimal set* of the problem L and is denoted by $\operatorname{Arg} L$. We can write

$$\operatorname{Arg} L = M \cap \{x \mid (c, x) = \alpha\}.$$

7.   A REFINED FORMULATION OF
THE FARKAS – MINKOWSKII THEOREM

Theorem 7.1. *If an inequality $(c, x) - \alpha \le 0$ is an implication of a system*

$$Ax \le b, \tag{7.1}$$

then for some $\lambda \ge 0$, $\lambda_0 \ge 0$ we have the identity

$$(c, x) - \alpha \overset{(x)}{\equiv} (Ax - b, \lambda) - \lambda_0. \tag{7.2}$$

In this case:

1) $\{a_j \mid \lambda_j > 0\}$ *are linearly independent;*

2) *the inequality $(c, x) - \alpha + \lambda_0 \le 0$ is a second-kind implication of system (7.1).*

Here $\lambda^T = [\lambda_1, \ldots, \lambda_m]$, $A = \begin{bmatrix} a_1 \\ \vdots \\ a_m \end{bmatrix}$, *i. e., a_j is the j-th row vector of the matrix A.*

Proof. Let $\tilde{\alpha} = \max_{x \in M}(c, x)$, where M is the set of solutions of system (7.1). Evidently, we have $\tilde{\alpha} \leq \alpha$. Applying the Farkas–Minkowski theorem to the inequality-implication $(c, x) \leq \tilde{\alpha}$ of second kind, we get

$$(c, x) - \tilde{\alpha} \overset{(x)}{\equiv} \sum_{j=1}^{m} \mu_j[(a_j, x) - b_j], \qquad (7.3)$$

where $\mu^T = [\mu_1, \ldots, \mu_m] \geq 0$. Suppose that $\bar{x} \in M \cap \{x \mid (c, x) = \tilde{\alpha}\}$. Relation (7.3) yields

$$0 = (c, \bar{x}) - \tilde{\alpha} = \sum_{j=1}^{m} \mu_j[(a_j, \bar{x}) - b_j] \leq 0. \qquad (7.4)$$

As follows from (7.4), if $(a_j, \bar{x}) - b_j < 0$, then $\mu_j = 0$. Consequently, if we set $J(\bar{x}) = \{j \mid (a_j, \bar{x}) = b_j\}$, then we can rewrite relation (7.3) as follows:

$$(c, x) - \tilde{\alpha} \overset{(x)}{\equiv} \sum_{j \in J(\bar{x})} \mu_j[(a_j, x) - b_j]. \qquad (7.5)$$

Since (7.5) yields

$$(c, x) = \sum_{j \in J(\bar{x})} \mu_j(a_j, x),$$

therefore, by Theorem 3.4, there exist $\lambda_j \geq 0$, $j \in J \subset J(\bar{x})$, such that

$$(c, x) \overset{(x)}{\equiv} \sum_{j \in J} \lambda_j(a_j, x),$$

where $\{a_j \mid j \in J\}$ are linearly independent. Let $\gamma = \sum_{j \in J} \lambda_j b_j$, then

$$(c, x) - \gamma \equiv \sum_{j \in J} \lambda_j[(a_j, x) - b_j]. \qquad (7.6)$$

Substituting $x = \bar{x}$ into (7.6) we obtain $(c, \bar{x}) - \gamma = 0$, i.e., $\gamma = \tilde{\alpha}$. Finally, we rewrite (7.6) in the form

$$(c, x) - \tilde{\alpha} \equiv \sum_{j \in J} \lambda_j[(a_j, x) - b_j] - \lambda_0$$

where $\lambda_0 = \alpha - \tilde{\alpha}$. If we set $\lambda_j = 0$ for $j \notin J$, then the above relation will be the desired relation (7.2) with properties 1) and 2). $\square$

Theorem 7.2. *The optimal set* $\text{Arg}\, L$ *of a solvable problem* (6.6) *is a certain k-face of the polyhedron* M.

Proof. The inequality $(c, x) \leq \alpha$ is a second-kind implication of $Ax \leq b \sim (a_j, x) \leq b_j$, $j = 1, \ldots, m$. Here $\alpha = \text{opt}\, L$. By Theorem 7.1,

$$(c, x) - \alpha \overset{(x)}{\equiv} \sum_J \lambda_j [(a_j, x) - b_j], \qquad (7.7)$$

where $\lambda_j > 0$, $\{a_j\}_{j \in J}$ are linearly independent. Now let $J = \{j_1, \ldots, j_k\}$. Consider the k-boundary system

$$\begin{cases} (a_{j_s}, x) = b_{j_s}, & s = 1, \ldots, k, \\ (a_j, x) \leq b_j, & j \neq j_1, \ldots, j_k, \end{cases} \qquad (7.8)$$

with the set of solutions $M(j_1, \ldots, j_k)$. Let us show that $\text{Arg}\, L = M(j_1, \ldots, j_k)$. Really, if $\bar{x} \in \text{Arg}\, L$, i.e., $\bar{x} \in M$ and $(c, \bar{x}) = \alpha$, then (7.7) yields

$$\sum_J \lambda_j [(a_j, \bar{x}) - b_j] = 0.$$

Since $(a_j, \bar{x}) - b_j \leq 0$ and $\lambda_j > 0$, therefore, $(a_j, \bar{x}) - b_j = 0$, $j \in J$, i.e., $\bar{x} \in M(j_1, \ldots, j_k)$; hence it follows that $\text{Arg}\, L \subset M(j_1, \ldots, j_k)$.

Conversely, let $\bar{x} \in M(j_1, \ldots, j_k)$, i.e., $\bar{x}$ satisfies system (7.8). Substituting $\bar{x}$ into (7.7), we obtain $(c, \bar{x}) - \alpha = 0$; i.e., $(c, \bar{x}) = \text{opt}\, L$. Hence it follows that $\bar{x} \in \text{Arg}\, L$, so $M(j_1, \ldots, j_k) \subset \text{Arg}\, L$. $\square$

8. CONDITIONS OF COMPATIBILITY OF A FINITE SYSTEM OF LINEAR INEQUALITIES

A condition of compatibility of a system

$$Ax \leq b \qquad (8.1)$$

is obtained from Lemma 6.1 which was of auxiliary character in Section 6. By the lemma, the incompatibility of system (8.1) is equivalent to the fact that the inequality $t \leq 0$ is an implication of the system

$$Ax \leq bt. \qquad (8.1_t)$$

By the Farkas – Minkowski theorem, we have

$$t \overset{(x,t)}{\equiv} (Ax - bt, \bar{\lambda})$$

for some $\bar{\lambda} \geq 0$. Hence for $t = 1$ we have

$$A^T \bar{\lambda} = 0, \qquad (b, \bar{\lambda}) = -1 \quad (< 0).$$

Therefore, incompatibility of system (8.1) is equivalent to solvability of the system

$$A^T \lambda = 0, \quad \lambda \geq 0, \quad (b, \lambda) < 0. \tag{8.2}$$

The above facts result in the following statement:

Theorem 8.1. *System (8.1) is compatible if and only if for each $\bar{\lambda} \geq 0$ such that $A^T \bar{\lambda} = 0$ the inequality $(b, \bar{\lambda}) \geq 0$ holds, i.e., the inequality $(b, \lambda) \geq 0$ is an implication of the system $A^T \lambda = 0, \lambda \geq 0$.*

Remark 8.1. Since for $\lambda = 0$ the implication

$$A^T \lambda = 0, \ \lambda \geq 0 \quad \Rightarrow \quad (b, \lambda) \geq 0$$

holds, we may consider only $\lambda \neq 0$ or, taking into account the normalization, $\sum_{j=1}^{m} \lambda_j = 1, \lambda \geq 0$. Therefore, Theorem 8.1 may be reformulated as follows:

System (8.1) is solvable if and only if the inequality $(b, \lambda) \geq 0$ is an implication of the system

$$A^T \lambda = 0, \quad \lambda \geq 0, \quad \sum_{j=1}^{m} \lambda_j = 1.$$

According to Theorem 4.2, the polyhedron of solutions of the system $A^T u = 0, u \geq 0$, which is a cone, has a finite nonnegative basis $\{s_k\}_{k \in J}$, i.e., a finite collection of solutions such that any other solution of the system is a nonnegative combination of the basis vectors. So, Theorem 8.1 can be reformulated as follows:

Theorem 8.2. *System (8.1) is compatible if and only if $(b, s_k) \geq 0$, $\forall k \in J$, where $\{s_k\}_{k \in J}$ is a nonnegative basis of the cone of solutions of the system $A^T u = 0, u \geq 0$.*

Now let us consider the problem of compatibility of a mixed system of linear inequalities

$$(a_j, x) \le b_j, \quad \forall j \in J_\le; \qquad (a_j, x) < b_j, \quad \forall j \in J_<, \qquad (8.3)$$

for $J_\le \cup J_< = \{1, \ldots, m\}$.

Theorem 8.3. *System* (8.3) *is compatible if and only if the inequality* $(b, u) \ge 0$ *is an implication of the system* $A^T u = 0$, $u \ge 0$. *In this case, if* $\sum_{j \in J_<} u_j > 0$, *then* $(b, u) > 0$.

Proof. *Necessity.* Evidently, system (8.3) is compatible if and only if the system

$$(a_j, x) \le b_j, \quad \forall j \in J_\le; \qquad (a_j, x) \le b_j - \varepsilon, \quad \forall j \in J_<, \qquad (8.4)$$

is compatible for some $\varepsilon > 0$. By Theorem 8.1, the inequality

$$\sum_{j \in J_\le} b_j u_j + \sum_{j \in J_<} (b_j - \varepsilon) u_j \ge 0$$

is an implication of the system $A^T u = 0$, $u \ge 0$, i. e., the inequality

$$\sum_{j=1}^m u_j b_j \ge \varepsilon \sum_{j \in J_<} u_j$$

holds for each u satisfying the above system. Hence the desired condition follows

$$\sum_{j=1}^m u_j b_j \begin{cases} \ge 0, & \sum_{J_<} u_j = 0, \\ > 0, & \sum_{J_<} u_j > 0. \end{cases}$$

Sufficiency. First we note that the theorem assumptions and the source Theorem 8.1 imply that system (8.1) is compatible. Now, proceeding from the condition

$$\left(A^T u = 0, \ u \ge 0 \right) \ \& \ \sum_{j \in J_<} u_j \ge 0 \quad \Rightarrow \quad (b, u) > 0,$$

we need to prove the compatibility of system (8.3), i. e., the system

$$(a_j, x) \le b_j, \quad j \in J_\le; \qquad (a_j, x) \le b_j - \varepsilon, \quad j \in J_<, \qquad (8.5)$$

for some $\varepsilon > 0$. If it is not so, i. e., if system (8.5) is incompatible for each $\varepsilon > 0$, then for each $\varepsilon > 0$ there exists a number $k(\varepsilon)$ such that

$$\sum_{j \in J_{\leq}} u_j^{k(\varepsilon)} b_j + \sum_{j \in J_{<}} u_j^{k(\varepsilon)} (b_j - \varepsilon) < 0, \tag{8.6}$$

where $\{u^k := [u_1^k, \dots, u_m^k]^T\}_k$ is a nonnegative basis of the cone of solutions of the system $A^T u = 0$, $u \geq 0$ (see Theorem 8.2); in our case we may think that $\sum_{j=1}^m u_j^k = 1$. Relation (8.6) can be rewritten as

$$(b, u^{k(\varepsilon)}) < \varepsilon \sum_{j \in J_{<}} u_j^{k(\varepsilon)}, \tag{8.7}$$

where $u^{k(\varepsilon)} = [u_1^{k(\varepsilon)}, \dots, u_m^{k(\varepsilon)}]^T$. Since the system $Ax \leq b$ is compatible, therefore, $(b, u^{k(\varepsilon)}) \geq 0$ (for otherwise we would obtain the inconsistent inequality $0 < 0$). But then, by the theorem assumption, $(b, u^{k(\varepsilon)}) > 0$ where the vector $u^{k(\varepsilon)}$ takes on a finite number of states from the set $\{u^k\}_k$. In view of the above, inequality (8.7) yields the relation

$$0 < \delta := \min_{k(\varepsilon)} (b, u^{k(\varepsilon)}) < \varepsilon \sum_{j \in J_{<}} u_j^{k(\varepsilon)} \leq \varepsilon$$

that is inconsistent for ε sufficiently small. Thus, the sufficiency is proved. $\square$

Now let us formulate Theorem 8.1 for the system

$$Ax \leq b, \qquad x \geq 0. \tag{8.8}$$

System (8.8) is compatible if and only if the inequality $(b, u) \geq 0$ is an implication of the system $A^T u \geq 0$, $u \geq 0$.

Really, rewrite system (8.8) in the form $\bar{A}x \leq \bar{b}$ where

$$\bar{A} = \begin{bmatrix} A \\ -E \end{bmatrix}, \qquad \bar{b} = \begin{bmatrix} b \\ 0 \end{bmatrix} \in \mathbb{E}_{m+n}, \qquad E = \begin{bmatrix} 1 & & 0 \\ & \ddots & \\ 0 & & 1 \end{bmatrix}$$

(E is the unit matrix). The compatibility condition for the system $\bar{A}x \leq \bar{b}$ is the implication

$$A^T u - Ev = 0, \qquad \bar{u} = \begin{bmatrix} u \\ v \end{bmatrix} \geq 0 \qquad \Rightarrow \qquad (\bar{b}, \bar{u}) \geq 0$$

or

$$A^T u \geq 0, \quad u \geq 0 \quad \Rightarrow \quad (b, u) \geq 0,$$

which was to be proved.

Remark 8.2. As follows from Theorem 8.1, the systems $Ax \leq b$ and $A^T u = 0$, $u \geq 0$, $(b, u) < 0$ have the following property: one and only one of them is compatible. Such systems are called *alternative* systems.

9.   THE CLEANING THEOREM

In this section, we discuss the compatibility of a system of linear inequalities

$$(a_j, x) \leq b_j, \quad j = 1, \ldots, m, \tag{9.1}$$

in relation to compatibility of some of its subsystems. Let $r > 0$ be the rank of system (9.1).

Theorem 9.1. *If all the subsystems of system (9.1) that consist of $r' \leq r + 1$ inequalities are compatible, then system (9.1) is also compatible.*

Proof. If under the theorem assumptions system (9.1) is incompatible, then, by Lemma 6.1, the inequality $t \leq 0$ is an implication of the system

$$(a_j, x) \leq b_j t, \quad j = 1, \ldots, m,$$

which, by Theorem 7.1 (Farkas – Minkowski), gives

$$t \overset{[x,t]}{\equiv} \sum_{j \in J} \lambda_j [(a_j, x) - b_j t], \qquad \lambda_j \geq 0, \quad |J| \leq r + 1. \tag{9.2}$$

Taking into account Lemma 6.1 and Theorem 7.1, we conclude that the subsystem $(a_j, x) \leq b$, $j \in J$, is incompatible, which contradicts the theorem hypothesis.   □

10.   SEPARABILITY OF NONINTERSECTING POLYHEDRONS

We suppose that the systems

$$Ax \leq b \quad \text{and} \quad Bx \leq d$$

are compatible and define the polyhedrons M and N, respectively. If $M \cap N = \varnothing$, then the hyperplane $H = \{x \mid (c, x) - \alpha = 0\}$ is called *strictly separating* M and N if

$$(c, x) - \alpha < 0, \quad \forall x \in M; \qquad (c, x) - \alpha > 0, \quad \forall x \in N. \tag{10.1}$$

Theorem 10.1. *If $M \cap N = \varnothing$, then there exists a hyperplane strictly separating them.*

Proof. By assumption, the combined system $Ax \leq b$, $Bx \leq d$ is incompatible. Then, by Theorem 8.1, there exist $\bar{u} \geq 0$ and $\bar{v} \geq 0$ such that

$$A^T \bar{u} + B^T \bar{v} = 0 \quad \& \quad (b, \bar{u}) + (d, \bar{v}) =: -\lambda_0 < 0. \tag{10.2}$$

The last relation yields the identity

$$(Ax - b, \bar{u}) - \frac{\lambda_0}{2} + (Bx - d, \bar{v}) - \frac{\lambda_0}{2} \overset{(x)}{\equiv} 0. \tag{10.3}$$

Set

$$l(x) := (c, x) - \alpha = (Ax - b, \bar{u}) - \frac{\lambda_0}{2}, \tag{10.4}$$

i. e., $c = A^T \bar{u}$, $\alpha = (b, \bar{u}) + \lambda_0 / 2$. Equality (10.3) yields

$$l(x) = -(Bx - d, \bar{v}) + \frac{\lambda_0}{2}. \tag{10.5}$$

Relations (10.4) and (10.5) yield (10.1), which was to be proved. $\square$

Remark 10.1. As is seen from (10.4), in order to construct a separating affine function $l(x)$ we need to know the vector $\bar{u} \geq 0$ that is determined (together with $\bar{v} \geq 0$) from the system

$$A^T u + B^T v = 0, \quad u \geq 0, \quad v \geq 0, \qquad (b, u) + (d, v) \leq -1.$$

Thus, the problem is reduced to solving a system of linear equations and inequalities.

Now let us consider a more general case: separation of a finite collection of nonintersecting polyhedral sets

$$M_i := \{x \mid A_i x \leq b_i\}, \quad i = 1, \dots, m. \tag{10.6}$$

Theorem 10.2. If $\bigcap_{(i)} M_i = \varnothing$, then $\exists P_i := \{x \mid (a_i, x) - \alpha_i \leq 0\}$:

$$\bigcap_{i=1}^{m} P_i = \varnothing \quad \& \quad P_i \supset M_i, \ \forall i.$$

Proof. Combining systems (10.6) yields an incompatible system of linear inequalities

$$A_i x \leq b_i, \quad i = 1, \ldots, m.$$

By Theorem 8.1,

$$\exists \bar{u}_i : \quad \sum_{i=1}^{m} A_i^T \bar{u}_i = 0 \quad \& \quad \sum_{i=1}^{m} (b_i, \bar{u}_i) = -\lambda_0 < 0. \tag{10.7}$$

Set $a_i := A_i^T \bar{u}_i$, $\alpha_i := (b_i, \bar{u}_i)$. Relations (10.7) are the condition of incompatibility of the system

$$(a_i, x) \leq \alpha_i, \quad i = 1, \ldots, m.$$

Therefore, we have

$$\bigcap_{i=1}^{m} \left(P_i = \{x \mid (a_i, x) \leq \alpha_i\} \right) = \varnothing.$$

On the other hand, since $(a_i, x) - \alpha_i = (A_i x - b_i, \bar{u}_i)$, it follows that $\bar{x} \in M_i$ $\Rightarrow (a_i, \bar{x}) \leq \alpha_i$, which means that $P_i \supset M_i, \ \forall i.$ $\qquad \square$

Remark 10.2. In the case $m = 2$, i.e., when we have two sets M_1 and M_2, the system of two inequalities $(a_1, x) \leq \alpha_1$, $(a_2, x) \leq \alpha_2$ in the above theorem will be incompatible. This means that the boundary hyperplanes H_1 and H_2 of the half-spaces P_1 and P_2 are parallel. In this case, for some $\lambda > 0$ we have $a_1 = -\lambda a_2$ and $\alpha_1 < -\lambda \alpha_2$. For each $\alpha \in (\alpha_1, -\lambda \alpha_2)$ the affine function $l(x) := (a_1, x) - \alpha$ will strictly separate the polyhedrons M_1 and M_2. In our case, we have $l(x) < 0, \ \forall x \in M_1$, and $l(x) > 0, \ \forall x \in M_2$.

Remark 10.3. As in Remark 10.1, the vectors $\bar{u}_i$, $i = 1, \ldots, m$, that form a_i and α_i, i.e., the separating affine functions $\{(a_i, x) - \alpha_i\}_1^m$, are found from the system of linear inequalities and equations:

$$\sum_{i=1}^{m} A_i^T u_i = 0, \quad u_i \geq 0, \quad i = 1, \ldots, m, \quad \sum_{i=1}^{m} (b_i, u_i) \leq -1.$$

11. THE FOURIER ELIMINATION METHOD

Here we consider the elimination method for system (11.1) which can be conceived as a certain analog of the Gauss elimination method. For the system

$$l_j(x) = (a_j, x) - b_j \le 0, \quad j = 1, \ldots, m, \tag{11.1}$$

we set $J_+ = \{j \mid a_{j1} > 0\}$, $J_- = \{j \mid a_{j1} < 0\}$, $J_0 = \{j \mid a_{j1} = 0\}$. For definiteness, below we consider elimination of the variable x_1. Dividing the inequalities of system (11.1) by $|a_{j1}|$ for $j \in J_+ \cup J_-$, we rewrite the system as follows:

$$\begin{cases} x_1 + \bar{l}_j(y) \le 0, & j \in J_+, \\ -x_1 + \bar{l}_i(y) \le 0, & i \in J_-, \\ \bar{l}_k(y) \le 0, & k \in J_0, \end{cases} \tag{11.2}$$

where

$$y^T = [x_2, \ldots, x_n]; \qquad \bar{l}_j = \frac{l_j(x)}{a_{j1}} - x_1, \quad j \in J_+;$$

$$\bar{l}_i(y) = \frac{l_i(y)}{|a_{i1}|} + x_1, \quad i \in J_-; \qquad \bar{l}_k(y) = l_k(x), \quad k \in J_0.$$

System (11.2) generates the system of linear inequalities

$$\begin{cases} \bar{l}_i(y) \le x_1 \le -\bar{l}_j(y), & i \in J_-, \quad j \in J_+, \\ \bar{l}_k(y) \le 0, & k \in J_0. \end{cases} \tag{11.3}$$

Eliminating the variable x_1 from (11.3) we obtain the system

$$\begin{cases} \bar{l}_i(y) + \bar{l}_j(y) \le 0, & (i,j) \in J_- \times J_+, \\ \bar{l}_k(y) \le 0, & k \in J_0. \end{cases} \tag{11.4}$$

The above system is the result of elimination of the variable x_1 from the initial system (11.1).

Systems (11.1) and (11.4) are connected by the following theorem:

Theorem 11.1. *System (11.1) is compatible if and only if system (11.4) is compatible. In this case:*

1. *If $\bar{x}^T = [\bar{x}_1, \bar{y}]$ is some solution of system (11.1), then $\bar{y}$ is a solution of system (11.4).*

   2. *If $\bar{y}$ is some solution of system* (11.4), *then there exists $\bar{x}_1$ such that $\bar{x}^T = [\bar{x}_1, \bar{y}]$ is a solution of system* (11.1).

   Proof. The proof is contained, in essence, in the very construction of system (11.4). The first part of the theorem is evident. Let us clarify statement 2. Let $\underline{l} = \max_{i \in J_-} \bar{l}_i(\bar{y})$, $\bar{l} = \min_{j \in J_+}[-\bar{l}_j(\bar{y})]$. Then (11.3) for $y = \bar{y}$ implies that the inequality $\underline{l} \leq \bar{l}$ is consistent. Taking an arbitrary $\bar{x}_1$ from the interval $\overline{\underline{l}, \bar{l}}$ we obtain the solution $\bar{x}^T = [\bar{x}_1, \bar{y}]$ of system (11.3), i. e., a solution of system (11.1), which was to be proved. $\square$

   Remark 11.1. If one of the sets J_+, J_- is empty and, moreover, $J_0 = \varnothing$, then system (11.4) will be called *empty*. We shall say, by definition, that each vector $y \in \mathbb{E}_{n-1}\{e_2, \ldots, e_n\}$ satisfies it. Here $e_i^T = [0, \ldots, 1, 0, \ldots, 0]$ is the unit basis vector of the space $\mathbb{X}_n$; $\mathbb{X}_{n-1}\{e_2, \ldots, e_n\}$ is the subspace of $\mathbb{X}_n$ stretched on the unit basis vectors $\{e_i\}_{i=2}^n$.

   Remark 11.2. Elimination of several variables from system (11.1) can be realized successively. This procedure with the variables x_i, $i \neq i_1, \ldots, i_s$, realizes the projection of M onto $\mathbb{X}_s\{e_{i_1}, \ldots, e_{i_s}\} \subset \mathbb{X}_n$.

   Corollary 11.1. *The set of solutions M_1 of system* (11.4) *is the algebraic projection of M onto $\mathbb{X}_{n-1}\{e_2, \ldots, e_n\} \subset \mathbb{X}_n$.*

   Corollary 11.2. *If for $i > 1$ we have $\inf_{x \in M} x_i > -\infty$ ($\sup_{x \in M} x_i < +\infty$), then $\min_{x \in M} x_i = \min_{y \in M_1} x_i$ ($\max_{x \in M} x_i = \max_{y \in M_1} x_i$).*

   Remark 11.3. The elimination procedure described above consists in combining the inequalities $l_j(x) \leq 0$, $j \in J_+$, with positive coefficients before x_1 and the inequalities $l_j(x) \leq 0$, $j \in J_-$, with negative coefficients before x_1. These inequalities are multiplied by positive numbers λ_j and μ_i such that $\lambda_j a_{j1} + \mu_i a_{i1} = 0$ and then summed up. Such numbers may be $\lambda_j = |a_{i1}|$, $\mu_i = a_{j1}$. Evidently, one may multiply only one of the inequalities $l_j(x) \leq 0$, $l_j(x) \leq 0$ by a positive number such that the zero coefficient before x_1 after their summation is ensured.

   Remark 11.4. The number of inequalities in system (11.4) is equal to $|J_+| \cdot |J_-| + |J_0|$. This shows that the number of inequalities in systems obtained in the result of elimination grows, and the growth is dramatic for a certain relation of signs before the variables being eliminated.

Let us mark system (11.4) by the symbol S_1. The system obtained after eliminating x_2 from the system S_1 will be denoted by the symbol S_2, etc. Having eliminated consecutively $x_1, \ldots, x_k$, we obtain the resulting system S_k. For $k = n - 1$, it will have only one variable x_n. Having found $\bar{x}_n$ from the last system, we substitute it into the system S_{n-2} to obtain a system with only one variable x_{n-2}. From this system we find $\bar{x}_{n-2}$ and so on. Finally, having come to the system S_1, we form the vector $\bar{y}^T = [\bar{x}_2, \ldots, \bar{x}_n]$. If we substitute it into the initial system (11.1) and find $\bar{x}_1$ from this system, then $\bar{x}^T = [\bar{x}, \bar{y}]$ will be one of solutions of (11.1).

On each step of the above sequential procedure (to find a vector $\bar{x}$) we solve a system of inequalities with one variable. Such a system is a simple object and we give it a special consideration. Simplifying the notation, we write a system of linear inequalities with one variable t in the form

$$\alpha_j t \leq \beta_j, \quad j = 1, \ldots, m. \tag{11.5}$$

Set

$$\underline{t} = \begin{cases} \max_{\alpha_j < 0} \beta_j / \alpha_j, & \{j \mid \alpha_j < 0\} \neq \varnothing, \\ -\infty, & \{j \mid \alpha_j < 0\} = \varnothing; \end{cases}$$

$$\bar{t} = \begin{cases} \min_{\alpha_j > 0} \beta_j / \alpha_j, & \{j \mid \alpha_j > 0\} \neq \varnothing, \\ +\infty, & \{j \mid \alpha_j > 0\} = \varnothing. \end{cases}$$

The condition $\underline{t} \leq \bar{t}$ is necessary and sufficient for system (11.5) to be compatible. For the set T of solutions of system (11.5) the following cases are possible:

1. T is the segment with the endpoints $\underline{t}$ and $\bar{t}$ if these numbers are finite.

2. $T = \{t \mid t \leq \bar{t}\}$ if $\bar{t} < +\infty$ and $\underline{t} = -\infty$, i.e., T is the half-axis with the right endpoint $\bar{t}$.

3. $T = \{t \mid \underline{t} \leq t\}$ if $\underline{t} > -\infty$ and $\bar{t} = +\infty$, i.e., T is the half-axis with the left endpoint $\underline{t}$.

In this classification, we have excluded the case when system (11.5) contains improper inequalities (this case corresponds to $\alpha_j = 0$).

Note that, if the problem is to find the maximum or minimum of αt with the restrictions (11.5), then

$$\max_{(11.5)} \alpha t = \begin{cases} \alpha \bar{t}, & \alpha > 0, \\ \alpha \underline{t}, & \alpha < 0; \end{cases} \qquad \min_{(11.5)} \alpha t = \begin{cases} \underline{t}, & \alpha > 0, \\ \bar{t}, & \alpha < 0. \end{cases}$$

Chapter 2.

Linear programming

---

12.   SETTING OF THE PROBLEM OF LINEAR PROGRAM- MING AND SOME ITS PROPERTIES

The problem of *linear programming* (LP) is understood as the problem to maximize (minimize) a linear functional (c, x) on the set of solutions of a system of linear inequalities, i. e., as the problem

$$\max\{(c, x) \mid (a_j, x) - b_j \leq 0, \ j = 1, \ldots, m\}. \tag{12.1}$$

Strictly speaking, the use of the operator max instead of sup in (12.1) should be clarified. If α is the finite least upper bound of the values of the functional (c, x) in the polyhedron M defined by the system of inequalities in (12.1), then, as was mentioned in Section 5 (Theorem 5.3), the inequality

$$(c, x) - \alpha \leq 0,$$

where $\alpha = \sup\{(c, x) \mid x \in M\}$, is a second-kind implication of this system. By Theorem 5.1, the set

$$\widetilde{M} = \{\, x \in M \mid (c, x) = \alpha\}$$

is non-empty; and in this case it defines those values x from M where α is attained.

The problem (12.1) is called *solvable* if $M \neq \varnothing$ and

$$\alpha = \sup\{(c, x) \mid x \in M\} < +\infty.$$

The sets M and $\widetilde{M}$ introduced for the problem (12.1) are called *admissible* and *optimal*, respectively; and vectors $x \in M$ and $\tilde{x} \in \widetilde{M}$ are called *admissible* and *optimal* vectors, respectively. The number α is called the *optimal value*; and $\{\tilde{x}, \tilde{\alpha}\}$ is called the *solution* of the problem (12.1).

If the problem (12.1) is solvable, then its optimal set $\widetilde{M}$ is evidently convex and polyhedral. Moreover, it is some k-face of the polyhedron M (Theorem 7.2).

Theorem 12.1. *Let* (12.1) *be a solvable LP problem in the space* $\mathbb{E}_n$, *and the rank of its system of constraints be equal to* n. *Then the optimal set* $\widetilde{M}$ *of this problem contains at least one vertex of the polyhedron* M *of its admissible vectors.*

Proof. The proof follows from the definition of a vertex of a polyhedron as its n-face and from Theorems 2.1 and 7.2.   □

Let us give one of the simplest conditions of solvability of the LP problem (12.1).

Theorem 12.2. *The problem* (12.1) *with* $M \neq \varnothing$ *is solvable if and only if for some* $u_j \geq 0$, $j = 1, \ldots, m$, *we have*

$$c = \sum_{j=1}^{m} u_j a_j. \tag{12.2}$$

Proof. Let the problem (12.1) be solvable and α be its optimal value. Then we have

$$(c, x) - \alpha \equiv \sum_{j=1}^{m} u_j [(a_j, x) - b_j]$$

(see formula (7.7)), which with $u_j = 0$ for $j \notin J$ yields relation (12.2).

Conversely, if (12.2) holds, then for each $y \in M$ we obtain

$$(c, y) = \sum_{j=1}^{m} u_j (a_j, y) \leq \sum_{j=1}^{m} u_j b_j = \gamma < +\infty,$$

i. e.,

$$\sup\{(c, y) \mid y \in M\} \leq \gamma < +\infty.$$

Therefore, the problem (12.1) is solvable.   □

13. ECONOMIC INTERPRETATION OF THE LINEAR PRO-GRAMMING PROBLEM

In further treatment we shall invoke meaningful economic interpretations. In this connection let us give several standard forms for systems of linear inequalities and LP problems. We shall use the following standard forms for systems of linear inequalities:

$$Ax \leq b, \tag{13.1}$$
$$Ax \leq b, \qquad x \geq 0, \tag{13.2}$$
$$Ax = b, \quad x \geq 0. \tag{13.3}$$

The corresponding forms for LP problems are

$$\max\{(c, x) \mid Ax \leq b\}, \tag{13.4}$$
$$\max\{(c, x) \mid Ax \leq b, \ x \geq 0\}, \tag{13.5}$$
$$\max\{(c, x) \mid Ax \leq b, \ x \geq 0\}. \tag{13.6}$$

For (13.1) and the LP problem in the form (13.4), the theorems proved below will have simpler form. The forms (13.2) and (13.5) are convenient for economic interpretation. The form (13.6) for the LP problem is called *canonical*. It is usually taken as initial form for expounding the direct simplex-method of solution of LP problems.

Systems (13.1)–(13.3) are equivalent in the sense of possibility of passing from one to another by application of the following elementary transformations:

1) $l(x) = 0 \xrightarrow{\sim} l(x) \leq 0, \ -l(x) \leq 0$;

2) $l(x) \leq 0 \xrightarrow{\sim} l(x) + x_{n+1} = 0, \ x_{n+1} \geq 0$;

3) if x is a free variable, then $x \xrightarrow{\sim} x = u - v, \ u \geq 0, \ v \geq 0$, i.e., one can pass from free variables to non-negative variables.

As for the equivalence of the settings (13.4)–(13.6), we need only take into account the possibility of the equivalent passage:

$$\max_{x \in M}(\min) f(x) \xrightarrow{\sim} -\min_{x \in M}(\max)[-f(x)].$$

Consider a problem in the form (13.5):

$$\max\{(c, x) \mid (a_j, x) \leq b_j, \ j = 1, \ldots, m, \ x \geq 0\}, \tag{13.7}$$

where $c = [c_1, \ldots, c_n]^T$, $a_j = [a_{j1}, \ldots, a_{jn}]$, $[a_{ji}]_{m,n} = A$. Let $p_i = [a_{1i}, \ldots, a_{mi}]^T$, i.e., p_i is the i-th column of the matrix A.

Suppose that there is some *production* $\mathcal{U}$ that *transforms* a system $b_1, \ldots, b_m$ of source *ingredients* which have certain *value* into a new system of valuable entities—the system of *products*. The terms 'production', 'value', 'ingredient', 'product' and 'evaluation' (as measure of worth; this term will appear below) will be used as starting terms.

The source ingredients being transformed may be fixed and current assets (buildings, equipment, transport, raw materials, electrical energy, etc.), natural resources (minerals, land, water of rivers and lakes, etc.), and labour resources which are classified by speciality and level of qualification.

Let us lay some finite collection of *technological methods* modelled by the vectors

$$p_i = [a_{1i}, \ldots, a_{mi}]^T \in \mathbb{E}_m, \quad i = 1, \ldots, n,$$

into the foundation of transformation of the ingredients. Here the coordinates of the vector p_i are interpreted as expenditures (inputs) of ingredients per unit *intensity* (for example, per unit of time) of using the i-th technological method. The intensity will be denoted by x_i ($x_i = 1$ is the unit intensity).

Suppose that the ingredient input and created value depend linearly on the vector of intensities $x = [x_1, \ldots, x_n]^T$. Under this assumption, the input of the j-th ingredient corresponding to the vector x is (a_j, x), $j = 1, \ldots, m$. The upper bounds for these inputs are set by the numbers b_j, therefore, the following inequalities should hold true:

$$(a_j, x) \le b_j, \quad j = 1, \ldots, m.$$

Since the vector x is the level of production, it should be non-negative: $x \ge 0$. Thus, we have obtained the system of constraints for the problem (13.5). The above discourse can be summarized in the scheme

$$\mathbb{E}_n \ni x \xrightarrow{\mathcal{U}} Ax = y \in \mathbb{E}_m,$$

where $x \ge 0$, $y \le b = [b_1, \ldots, b_m]^T$.

Suppose that c_i is the value of the final results of application of the i-th method of production with unit intensity $x_i = 1$. Then

$$(c, x) = \sum_{i=1}^n c_i x_i$$

is the total value of production defined by the level of production x.

Therefore, problem (13.5) is to determine such level of production (the intensity vector) from a certain class defined by linear restrictions which attains for (c, x) the maximal value.

The following modification for problem (13.7) is possible also. Namely, we suppose that x_i is the production level of ith product, c_i is the value of unity of ith product. Then (c, x) is the total profit which must be maximized.

We shall consider now the useful interpretation of the LP problem in the form

$$\min\{(c, x) \mid Ax \leq b, \ Bx \geq d, \ x \geq 0\}. \tag{13.8}$$

The technological methods $i = 1, \ldots, n$ and their intensities will be the basis of the setting; c_i are the labour inputs correspondent to the unit intensity $x_i = 1$; b is the resource vector; p_i (ith column of the matrix A) is the vector of resource consuming correspondent to the unit intensity; h_i (ith column of the matrix B) is the vector of production correspondent to the unit intensity; d is a fixed production level. The model (13.8) in these terms is written as follows: find the production level $x \geq 0$ satisfying the resource restrictions $Ax \leq b$ and the conditions of necessary level of production $Bx \geq d$ which provides the minimal labour input.

In essence, model (13.8) and its interpretation (taking into account the possibility of widening of the terms) satisfies the general economic level. If we set $M = \{x \geq 0 \mid Ax \leq b\}$, $N = \{x \geq 0 \mid Bx \geq d\}$, the situation may be pictured (see Figure 1). In this figure $S = M \cap N \neq \varnothing$, $\bar{x}$ is the vector of optimal production level corresponding to the vector $\bar{c}$ chosen in the figure. The polyhedrons M and N are the domain of resource possibilities and requirements. In practice (generally speaking) the polyhedrons M and N expressed mathematically as $S = M \cap N = \varnothing$, This situation will be considered in details in Chapter 3.

The generalization of model (13.8) in the sense of restrictions may be the system of inequalities

$$A_j x \begin{pmatrix} \leq \\ \geq \\ = \end{pmatrix} b^j, \quad j = 1, \ldots, k, \quad x \geq 0, \tag{13.9}$$

where $A_j x$ and b^j are combined by one of the three signs $\leq$, $\geq$, or $=$. This gives us the complete possibility of taking into account all the economic factors correspondent to the technology process. Different variants of model (13.8) with restrictions (13.9) give us various classic models of mathematical economy. Various subsystems in (13.9) may be inconsistent when we combine them (see Chapter 3).

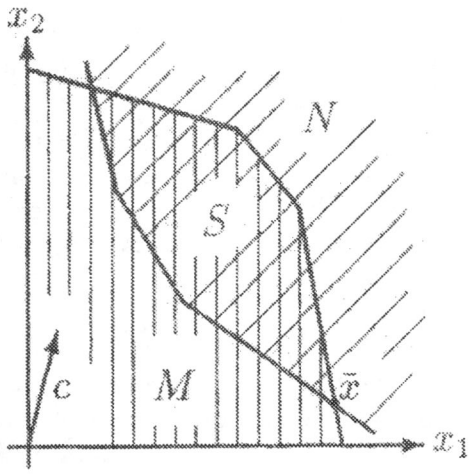

Figure 1

14.   DUALITY: INFORMATIVE APPROACH

Duality plays fundamental role both in mathematical programming and in mathematical in general. It generates the variety of constructive methods for analysis of mathematical models, for construction the effective algorithms for solution of problems.

If $\mathcal{M}$ is a certain initial mathematical object (model), then the dual object $\mathcal{M}^*$, generally speaking, is considered as an *external* object with respect to $\mathcal{M}$ (the object of observation of the object $\mathcal{M}$). The duality, of course, has its own specific character depending on the concrete essence defined by concrete mathematical theory (algebra, functional analysis, convex analysis, theory of optimal control and so on). This role of duality allows to comprehend mathematics in more complete sense.

The essence of duality in mathematical programming consists in considering the problem C^* parallel with the problem C which is formed following certain rules. This problems C and C^* is called *dual* problem. The problems C and C^* are connected by mathematical relations. These relations allow us to obtain, for example, the estimates of efficiency of all the parameters which form the problem C. As a result we reduce the solution of optimization problem to a certain system of linear inequalities; we form the optimality condition in more convenient form; we can estimate the speed of convergence of iterative processes for the problem C and so on.

If the LP problem is a result of modelling a certain economic (production) problem, then the duality and the information generated by this duality allow us to analyze the situation (the object we model) expressing the facts numerically. Such analysis us called the *economic-mathematical analysis*.

If we take the problem (13.5) as the initial LP problem, i. e.,

$$L: \ \max\{(c, x) \mid Ax \leq b, \ x \geq 0\},$$

then the *dual* problem (formally) will be as follows:

$$L^*: \ \min\{(b, u) \mid A^T u \geq c, \ u \geq 0\}.$$

Below we shall formulate some meaningful approaches which for the problem L^*.

14.1. The two-person game with zero sum

The two-person game is defined by 1) the collection of players (the 1st player, the 2nd player); 2) the sets X and U of strategies x and u respectively of 1st and 2nd players; 3) the payoff functions $F_1(x, u)$ and $F_2(x, u)$ which express the gains of the players for independent choice of strategies $x \in X$ and $u \in U$. If $F_1(x, u) + F_2(x, u) = 0$, which means that the gain of one player is equal to the loss of the second one, the gain is called the zero sum gain. In this case we can operate only with the payoff function of the 1st player $F_1(x, u) := F(x, u)$. This function is the loss function of the 2nd player. One of the most principles of such game is the principle of *guaranteed result*. It denotes that we search for the strategies which provide to the players the guaranteed gains. This approach may be formalized as follows. If the 1st player fixes the strategy $x \in X$, its guaranteed gain will be $E_1(x) = \min_{u \in U} F(x, u)$. Then we maximaze this gain

$$\max_{x \in X} E_1(x) = \max_{x \in X} \min_{u \in U} F(x, u).$$

For the 2nd player such approach gives

$$\max_{u \in U} \min_{x \in X} [-F(x, u)] = -\min_{u \in U} \max_{x \in X} F(x, u).$$

Thus, we have obtained the two problems

$$\max_{x \in X} \min_{u \in U} F(x, u) =: F_1, \qquad (14.1)$$

$$\min_{u \in U} \max_{x \in X} F(x, u) =: F_2, \qquad (14.2)$$

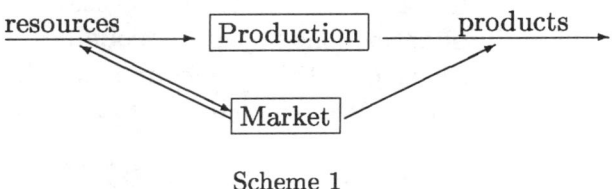

Scheme 1

If $F_1 = F_2$ $(=: F_0) = F(\bar{x}, \bar{u})$, $\bar{x} \in X$, $\bar{u} \in U$, then the number F_0 is called *the value of the game*; $\bar{x}$ and $\bar{u}$ are the optimal strategies of the 1st and 2nd players respectively (equilibrium strategies). Generally speaking F_1 and F_2 may not coincide. Problems (14.1) and (14.2) we call *formalization* of the two-person game with zero sum on the principle of the guaranteed result.

14.2. The model "Production – Market"

The problem L may be considered as the model of Production (taking into account the economic interpretation). We make now widening of this construction introducing the Market. As a result we obtain the system Production – Market

We consider now Production as the 1st player with the set of strategies $X = \{x \geq 0\} \subset \mathbb{E}_n$ and Market as the 2nd player with the functions: 1) input (buying) of output of Production with fixed values c_i, $i = 1, \dots, n$ and 2) input or sale of resources in the system Production – Market. The set of strategies of the Market will be the collection of price of resources unit with the number j, $j = 1, \dots, m$. The payoff function is as follows

$$F(x, u) = (c, x) - (Ax - b, u) \ \Big(= (c, x) - \sum_{j=1}^{m} u_j[(a_j, x) - b_j]\Big). \qquad (14.3)$$

It is called the Lagrange function with respect to the problem L. This function, evidently, has the sense of the gain (income) of the 1st player as a result of the two strategies $x \in X$, $u \in U$. This is the result of change operations with output and resources: (c, x) is the income of x_i (with fixed prices c_i), $i = 1, \dots, n$. If $(a_j, x) - b_j > 0$, then $u_j[(a_j, x) - b_j]$ is the charge of the 1st player to 2nd player for buying jth resource. The quantity of this resource is $(a_j, x) - b_j$; the price is u_j. If $(a_j, x) - b_j < 0$ then the term $-u_j[(a_j, x) - b_j] = u_j[b_j - (a_j, x)]$ is red as income of the 1st player when he sales the excess of jth product (the quantity is $b_j - (a_j, x)$; the price is u_j, $j = 1, \dots, m$). Therefore, the total sum $-\sum_{j=1}^{m} u_j[(a_j, x) - b_j] =$

$-(Ax - b, u)$ is the income of the 1st player when he sales the excess and buys the resources which are needed for the production level x; $F(x, u)$ is the total income of the 1st player.

The principle of the guaranteed result applied to the game formulated above leads to the problems

$$\max_{x \geq 0} \min_{u \geq 0}[(c, x) - (Ax - b, u)], \tag{14.4}$$

$$\min_{u \geq 0} \max_{x \geq 0}[(c, x) - (Ax - b, u)], \tag{14.5}$$

All the definitions formulated for problems (14.1) and (14.2) can be applied to these problems. In this situation, the optimal strategy of the 1st player (Production) is naturally called by the *optimal plan* and the optimal strategy of the 2nd player (Market) is called by the vector of *optimal prices* for resources.

In Subsection 14.5 the notions of optimal plan in the sense of optimal strategy and the optimal plan for the problem L coincide.

14.3. The play approach to duality

The qualitative approach to duality, i.e. the approach to the problem L^* dual to L can be realized by introducing the following equivalencies $\overset{\sim}{\rightarrow}$

$$(14.4) \overset{\sim}{\rightarrow} L, \tag{14.6}$$

$$(14.5) \overset{\sim}{\rightarrow} L^*,. \tag{14.7}$$

Here $\overset{\sim}{\rightarrow}$ denotes the equivalent transformation. Now we consider (14.6) and the internal operation in (14.4): $\min_{u \geq 0} F(x, u)$. We have

$$\min_{u \geq 0} F(x, u) = \min_{u \geq 0}[(c, x) - (Ax - b, u)] = \begin{cases} (c, x), & \text{if } Ax \leq b; \\ -\infty, & \text{if } Ax \not\leq b \end{cases}$$

(here $x \geq 0$). Hence it follows

$$\max_{x \geq 0} \min_{u \geq 0} F(x, u) = \max_{x \geq 0}\{(c, x) \mid Ax \leq b\},$$

i.e., (14.6) holds. In order to install (14.7) we transform the Lagrange function

$$F(x, u) = (c, x) - (Ax - b, u) = (b, u) + (-A^T u + c, x). \tag{14.8}$$

We consider now the internal operation in (14.5), i. e., $\max_{x \geq 0} F(x, u)$. We obtain

$$\max_{x \geq 0} F(x, u) = \max_{x \geq 0}[(b, u) + (-A^T u + c, x)] = \begin{cases} (b, u), & \text{if } A^T u \geq c; \\ +\infty, & \text{if } A^T u \not\geq c \end{cases}$$

(here $u \geq 0$). Hence it follows

$$\min_{u \geq 0} \max_{x \geq 0} F(x, u) = \min_{u \geq 0}\{(b, u) \mid A^T u \geq c\}.$$

So, the problem L^* dual to L we obtain as the equivalent of play problem (14.5) in the framework of the play model Production – Market. In this case, (14.5) (and, therefore, L^*) is the problem of construction of optimal (equilibrium) prices of resources. At the same time we have obtained the equivalence of the problem L and play problem (14.4). This, in particular, denotes that $\bar{x}$ is optimal for L if $\bar{x}$ the optimal strategy of game (14.4), (14.5). Conversely, if $\bar{x}$ is an optimal solution (optimal plan) of the problem L, then, for a certain $\bar{u} \geq 0$, the pair $[\bar{x}, \bar{u}]$ will determine the optimal strategies of the players.

14.4. The approach to duality where x and u in the Lagrange function are renamed

The Lagrange function had arisen first in researches in analytic mechanics (Lagrange, Fourier, Farkas, Ostrogradskii *et al.*) in the problem of equilibrium of mechanical systems in the forms of equations and inequalities. The equilibrium in mechanical systems is connected with the state correspondent with the minimum of potential energy.

The Lagrange function can be introduced in every problem connected with conditional extremum. In this case we have to fulfill a certain rule of such correspondence. We shall write these forms (direct and inverse) for the max and min problems in the two forms:

$$P_1 : \left. \begin{array}{l} \max f(x) \\ f_j(x) \leq 0, \ j = 1, \ldots, m, \\ x \geq 0 \end{array} \right\} \rightleftarrows$$

$$\rightleftarrows F_1(x, u) = f(x) - \sum_{j=1}^{m} u_j f_j(x), \quad u_j \geq 0, \ \forall j,$$

$$P_2: \quad \min f(x)$$
$$\left.\begin{array}{l} f_j(x) \le 0, \ j = 1, \ldots, m, \\ x \ge 0 \end{array}\right\} \rightleftarrows$$

$$\rightleftarrows F_2(x, u) = f(x) + \sum_{j=1}^{m} u_j f_j(x), \quad u_j \ge 0, \ \forall j.$$

The inverse correspondence $\leftarrow$ is ambiguous since we may rewrite $F_1(x, u)$ and $F_2(x, u)$ in the form

$$F_1(x, u) = f(x) + \sum_{j=1}^{m} u_j[-f_j(x)],$$

$$F_2(x, u) = f(x) - \sum_{j=1}^{m} u_j[-f_j(x)].$$

In this case the following relations holds

$$F_1(x, u) \longrightarrow \left\{ \begin{array}{l} P_3: \quad \min f(x) \\ \quad -f_j(x) \le 0, \ j = 1, \ldots, m, \\ \quad x \ge 0 \end{array} \right.$$

and

$$F_2(x, u) \longrightarrow \left\{ \begin{array}{l} P_4: \quad \max f(x) \\ \quad -f_j(x) \le 0, \ j = 1, \ldots, m, \\ \quad x \ge 0. \end{array} \right.$$

The scheme of these correspondences does not contradicts to the mechanical sense of the Lagrange function. This scheme can be represented briefly as follows:

$$P_1 \rightleftarrows F_1 \longrightarrow P_3,$$
$$P_2 \rightleftarrows F_2 \longrightarrow P_4. \tag{14.9}$$

We apply it now to the problem L and to its Lagrange function. Equality (14.8) allows to consider the Lagrange function $F(x, u)$ in either case. Namely, we consider u as a variable and x as a vector of Lagrange multipliers. Then, following the correspondence $F_1 \longrightarrow P_3$ we shall have

$$L \longrightarrow F(x, u) \longrightarrow \min\{(b, u) \mid A^T u \ge c, \ u \ge 0\},$$

i. e., the problem L^* had arized as the result of the correspondence $F_1 \longrightarrow P_3$, where the variables x and u in the Lagrange function are changed their places. The complete form of the scheme (14.9) is as follows

$$
\begin{array}{ccc}
\boxed{\begin{array}{c} \max(c,x) \\ Ax \le b,\ x \ge 0 \end{array}} & \xrightarrow{\ \ (*)\ \ } & \boxed{\begin{array}{c} \min(b,u) \\ A^T u \ge c,\ u \ge 0 \end{array}} \\
\downarrow & & \uparrow \\
(c,x) - (Ax - b, u) & = & (b,u) + (c - A^T u, x) \\
\| & & \| \\
(c,x) + (b - Ax, u) & = & (b,u) - (A^T u - c, x) \\
\downarrow & & \uparrow \\
\boxed{\begin{array}{c} \min(c,x) \\ Ax \ge b,\ x \ge 0 \end{array}} & \xrightarrow{\ \ (*)\ \ } & \boxed{\begin{array}{c} \max(b,u) \\ A^T u \le c,\ u \ge 0 \end{array}}
\end{array}
$$

Note the transform rule $(*)$ to the dual problem in the upper row of the scheme is consistent with the transform rule $(*)$ in the lower row. This will be verified in Section 15.

14.5. Duality which follows from the analysis of the Farkas – Minkowski theorem

We suppose that α is an optimal solution of the solvable problem

$$\max\{(c,x) \mid Ax \le b,\ x \ge 0\}, \tag{14.10}$$

i. e., $\alpha = (c, \bar{x})$, $\bar{x} \in \mathrm{Arg}\,(14.10)$. Then the inequality $(c,x) \le \alpha$ is the second order implication of its restriction system. By the Farkas – Minkowski theorem 5.2 and the remarks to this theorem we have

$$(c,x) - \alpha = (Ax - b, \bar{u}) - (\bar{v}, x) \tag{14.11}$$

for certain $\bar{u} \ge 0$, $\bar{v} \ge 0$ and for all x.

Relation (14.11) yields the two relations

$$c - A^T \bar{u} = -\bar{v} \le 0, \tag{14.12}$$
$$(c, \bar{x}) = (b, \bar{u}). \tag{14.13}$$

The first of them gives

$$M^* := \{u \ge 0 \mid A^T u \ge c\} \ne \varnothing.$$

As for $x \in M := \{x \geq 0 \mid Ax \leq b\}$ and $u \in M^*$ we have inequality $(c, x) \leq (b, u)$:

$$(c, x) \leq (A^T u, x) = (Ax, u) \leq (b, u), \qquad (14.14)$$

then (14.13) yields $(b, \bar{u}) = \min_{u \in M^*}(b, u)$, i.e., $\bar{u} \in \text{Arg} \min_{u \in M^*}(b, u)$. Thus, we have obtained from relation (14.11) the following:

1) the system of linear inequalities $A^T u \geq c$, $u \geq 0$, which defines the polyhedron M^*;

2) the problem $L^* : \min\{(b, u) \mid A^T u \geq c, \ u \geq 0\}$ which is dual to the problem;

3) the equality $(c, \bar{x}) = (b, \bar{u})$.

This is the duality theorem for the LP problem (see Section 15).

14.6. Construction of the dual problem on the basis of thermo-dynamical axiomatics

We shall take the following basic scheme

$$\underset{b}{\xrightarrow{\text{resources}}} \boxed{\text{Production}} \underset{x}{\xrightarrow{\text{product}}}$$

On the scheme we see the Transformer (pictured as Production) of the input (the resource vector b) into the output (the production vector x). In form (14.10) of the LP problem the vector of initial ingredients (resources) b has no price; however, the vector x has the price system. The system of prices may be considered more widely, namely, as the system of energy measurements. This system denotes that every energy transformation has efficiency coefficient not greater than unity. In our situation this denotes that the total energy estimate of the input is not less than the analogous estimate of the output. If c is considered as the vector of energy measurements of the production x and $u \in \mathbb{E}_n^+$ is unknown vector of energy measurements of initial ingredients then the above axiom can be written as follows

$$(h, u) \geq (c, x).$$

In particular, suppose $h = h_i$ is the ith column of the matrix A which expresses the expenditure of ingredients for the unity of ith production. We suppose that it is evaluated by c_i. Then the inequality takes the form

$$(h_i, u) \geq c_i.$$

The properties imposed on the Transformer we formulate in the form of the following axioms:

1) the total estimate of ingredients we use is depending linearly on the vector of their estimates u_j, $j = 1, \ldots, m$;

2) the total estimate of the ingredients we use in ith technological method on the unity production is not less than c_i which is the price of the unity of this production;

3) the optimal estimates of ingredients are derived from the condition of the maximum of the efficiency coefficient (EC) of the Transformer.

The conditions 1) and 2) yield the system of linear restrictions

$$(h_i, u) = \sum_{j=1}^{m} a_{ji} u_j \geq c_i, \quad i = 1, \ldots, n, \tag{14.15}$$

with the polyhedron M^*. The condition 3) formulates the problem

$$\mathrm{EC} = \frac{\text{output}}{\text{input}} = \frac{(c, x)}{(b, u)} \longrightarrow \max;$$

in this case, evidently, $x \in M$, $u \in M^*$. Hence it follows that

$$\mathrm{EC}_{opt} = \max \frac{\{(c, x) \mid x \in M\}}{\{(b, u) \mid u \in M^*\}} = \frac{\max\{(c, x) \mid x \in M\}}{\min\{(b, u) \mid u \in M^*\}}.$$

The numerator of this ratio is the problem L modelling the Production; the denominator is the problem L^* dual to L modelling the problem of determination of optimal estimates (in the energy sense) of initial ingredients.

The term energy estimate used for the problem L^* has purely associative sense and in the framework of our setting this term has no other sense.

15.   THE DUALITY THEOREM

We shall formulate now the basic statements which connect the problems

$$L : \quad \max\{(c, x) \mid Ax \leq b, \ x \geq 0\} \tag{15.1}$$

and

$$L^* : \quad \min\{(b, u) \mid A^T u \geq c, \ u \geq 0\}. \tag{15.1*}$$

First we introduce the standard notations: $\operatorname{Arg} L$ is the optimal set of the problem L; $\arg L$ is the concrete element from $\operatorname{Arg} L$; $\operatorname{opt} L$ is the optimal value of the problem L. If S is the symbol for notation of a certain system of linear inequalities; $\operatorname{Arg} S$ is the set of solutions of this system. The following relations are evident:

$$\operatorname{Arg} L = M \cap \{x \mid (c, x) = \operatorname{opt} L\},$$
$$\operatorname{Arg} L^* = M^* \cap \{u \mid (b, u) = \operatorname{opt} L^*\},$$

where $M = \{x \geq 0 \mid Ax \leq b\}$, $M^* = \{u \geq 0 \mid A^T u \geq c\}$.

Theorem 15.1. *The following inequality holds:*

$$(c, x) \leq (b, u), \qquad \forall x \in M, \quad \forall u \in M^*$$

(see (14.14)).

Corollary 15.1. *If $(c, \bar{x}) = (b, \bar{u})$, $\bar{x} \in M$, $\bar{u} \in M^*$, then $\bar{x} \in \operatorname{Arg} L$, $\bar{u} \in \operatorname{Arg} L^*$.*

Theorem 15.2 [the duality theorem]. *If the problem L is solvable, then the problem L^* is solvable also; moreover, their optimal values coincide.*

The proof, in essence, contains in relations (14.12) and (14.13) which are the corollaries of the Farkas–Minkowski theorem. If we apply this theorem in the form of relation (14.11) we obtain the vector $\bar{u} \in M^*$ such that $(c, \bar{x}) = (b, \bar{u})$, where $\bar{x}$ is the optimal vector for L. By Corollary 15.1 this finishes the proof.

The symbol $\overset{(*)}{\to}$ will denote the *transformation* from the problem L written in form (15.1) to the dual problem L^*. If the initial LP problem is written in another form, we must rewrite this problem in the form (15.1) and then apply the rule $(*)$ and pass to the dual problem.

Theorem 15.3. *The dual problem to the dual problem L^* coincide with L: $(L^*)^* = L$.*

Proof. The problem L^* we rewrite in the form (15.1):

$$-\max\{(-b, u) \mid (-A^T)u \leq -c, \ u \geq 0\}.$$

Applying the rule $(*)$ to this problem we obtain

$$-\min\{(-c, x) \mid (-A^T)^T x \geq -b, \ x \geq 0\},$$

which means that

$$\max\{(c, x) \mid A^T x \leq b, \ x \geq 0\}.$$

□

Remark 15.1. The reciprocity rule proved above is universal, i.e., it does not depend on the form the initial problem.

Remark 15.2. Theorem 15.2, taking into account the reciprocity rule can be formulated as follows: *if the problem L^* is solvable, then the problem L is solvable also and their optimal values coincide.*

Theorem 15.2 yields the following statement.

Theorem 15.4. *The problem L is solvable if and only if $M \neq \varnothing$, $M^* \neq \varnothing$.*

The schemes of passing to the dual problems for (13.4) and (13.6):

$$\max\{(c, x) \mid Ax \leq b\} \xrightarrow{(*)} \min\{(b, u) \mid A^T u = c, \ u \geq 0\},$$

$$\min\{(c, x) \mid Ax = b, \ x \geq 0\} \xrightarrow{(*)} \max\{(b, u) \mid A^T u \leq c\}.$$

These schemes are mutually dual.

The following system of linear inequalities we shall consider together with problems (15.1) and (15.1*):

$$S: \left. \begin{array}{l} Ax \leq b, \quad A^T u \geq c, \\ x \geq 0, \quad u \geq 0, \\ (c, x) \geq (b, u). \end{array} \right\} \tag{15.2}$$

The problem of determining at least one solution of this system is called *the symmetric problem.*

The below theorem establishes the connections between the problems L and L^* in the forms (15.1) and (15.1*) and the system S.

Theorem 15.5. *The problem L and the system S are solvable (or unsolvable) simultaneously. If they are solvable, then*

$$\operatorname{Arg} S = \operatorname{Arg} L \times \operatorname{Arg} L^*,$$

which means that the set of solutions of the system S is the Cartesian product of the optimal sets of the problems L and L^*.

The proof follows from Corollary 15.1 and the duality theorem 15.2.

Remark 15.3. The following systems S correspond to problems (13.4) and (13.6) respectively

$$\left.\begin{array}{r}
Ax \leq b, \\
A^T u = c, \\
u \geq 0, \\
(c, x) \geq (b, u);
\end{array}\right\} \qquad (15.3)$$

$$\left.\begin{array}{r}
Ax = b, \\
x \geq 0, \\
A^T u \leq c, \\
(c, x) \leq (b, u).
\end{array}\right\} \qquad (15.4)$$

Now we shall give the scheme of forming the dual problem L^* if the problem L is given in the general form

$$L: \quad \max(c, x) \qquad \overset{(*)}{\underset{\longleftarrow}{\longrightarrow}} \qquad L^*: \quad \min(b, u)$$

with the restrictions $\qquad\qquad\qquad$ with the restrictions

$$Ax \left\{\begin{array}{ll}
\leq b_j, & j \in J_\leq \\
\geq b_j, & j \in J_\geq \\
= b_j, & j \in J_0
\end{array}\right. \qquad\qquad A^T u \left\{\begin{array}{ll}
\leq c_i, & i \in I_\leq \\
\geq c_i, & i \in I_\geq \\
= c_i, & i \in I_0
\end{array}\right.$$

$$\begin{array}{ll}
x_i \geq 0, & i \in I_\geq \\
x_i \leq 0, & i \in I_\leq \\
x_i \text{ is free}, & i \in I_0
\end{array} \qquad\qquad \begin{array}{ll}
u_j \geq 0, & j \in J_\leq \\
u_j \leq 0, & j \in J_\geq \\
u_j \text{ is free}, & j \in J_0.
\end{array}$$

All we were talking above about mutual properties of the dual LP problem is similar for this pair of problems $\{L, L^*\}$.

In conclusion we shall make some remarks on the problem of duality. If we consider the play approach to duality, i.e., to the dual problem L^* (see Section 14.3), the variable vector $u \geq 0$ is the vector of admissible prices for resources in the scheme "Production – Market". The optimal price vector $\bar{u} \geq 0$, taking into account equivalence (14.7) is the optimal vector of the problem L^*. Thus, the problem L as the production model with the fixed prices is completed by the model of determining the optimal prices (dual estimates) for resources.

Taking into account the duality approach stated in Subsection 14.6, we have

$$\text{EC}_{opt} = \frac{(c, \bar{x})}{(b, \bar{u})},$$

where $\bar{x} \in \text{Arg}\,L$, $\bar{u} \in \text{Arg}\,L^*$. By the duality theorem, we have $\text{EC}_{opt} = 1$ $((c, \bar{x}) = (b, \bar{u}))$. This means that if $\bar{x}$ is the optimal plan of Production and $\bar{u}$ is the vector of optimal prices for resources, the scheme pictured in pg. 59 is working with the efficiency coefficient equal to unity. Note that we shall not give deeper sense to this fact.

16.   THE OPTIMALITY CONDITIONS

We shall formulate below the conditions (of various forms) of optimality for admissible vector of the *LP* problem.

   Theorem 16.1. *An admissible vector $\bar{x}$ of problem* (13.4) *is optimal if and only if for certain*

$$\bar{u}_j \geq 0, \quad j \in J(\bar{x}) = \{j \mid (a_j, \bar{x}) = b_j\}$$

we have the relation

$$c = \sum_{j \in J(\bar{x})} u_j a_j^T. \tag{16.1}$$

   Proof. We suppose $\tilde{x}$ the optimal vector of problem (13.4). Then, taking into account Theorems 5.2 and 5.3, for certain $\tilde{u}_j \geq 0$, $j = 1, \ldots, m$, we have

$$(c, x) - (c, \tilde{x}) \overset{(x)}{\equiv} \sum_{j=1}^{m} \tilde{u}_j[(a_j, x) - b_j]. \tag{16.2}$$

This relation yields for $x = \tilde{x}$

$$\sum_{j=1}^{m} \tilde{u}_j[(a_j, \tilde{x}) - b_j] = 0.$$

Taking into account that all the terms are non-positive, we have

$$\tilde{u}_j[(a_j, \tilde{x}) - b_j] = 0, \qquad j = 1, \ldots, m.$$

Therefore, if $(a_j, \tilde{x}) - b_j < 0$, then $\tilde{u}_j = 0$. This fact allows us to re-place (16.2) by

$$(c, x) - (c, \tilde{x}) \equiv \sum_{j \in J(\tilde{x})} \tilde{u}_j[(a_j, x) - b_j],$$

whence follows relation (16.1).

Conversely, we suppose that for the optimal vector $\bar{x}$ of problem (13.4) the relation (16.1) holds. We shall prove that $\bar{x}$ is the optimal vector. If y is an arbitrary admissible vector for problem (13.4), then, taking into account condition (16.1), we have

$$(c, y) = \sum_{j \in J(\bar{x})} \bar{u}_j(a_j, y)$$

$$\leq \sum_{j \in J(\bar{x})} \bar{u}_j b_j = \sum_{j \in J(\bar{x})} \bar{u}_j(a_j, \bar{x}) = (c, \bar{x}).$$

Therefore, $\bar{x}$ is the optimal vector for problem (13.4). $\square$

Theorem 16.2. *Admissible vectors $\bar{x}$ and $\bar{u}$ for problems (15.1) and (15.1*) are optimal if and only if*

$$\bar{u}^T[A\bar{x} - b] = 0, \qquad \bar{x}^T[A^T\bar{u} - c] = 0. \tag{16.3}$$

Proof. If $\bar{x}$ is an optimal vector of problem (15.1) and $\bar{u}$ is an optimal vector of problem (15.1*), then, by Theorem 15.2, we have $\bar{u}^T b = \bar{x}^T c$. Hence, taking into account the evident equality $\bar{u}^T A\bar{x} = \bar{x}^T A^T \bar{u}$, we have

$$\bar{u}^T[A\bar{x} - b] + \bar{x}^T[c - A^T\bar{u}] = 0.$$

In this relation both terms are non-negative; therefore, (16.3) holds true. Conversely, (16.3) yields $\bar{u}^T b = \bar{x}^T c$. As $\bar{x}$ and $\bar{u}$ are admissible for problems (15.1) and (15.1*) respectively, then, by Corollary 15.1, the vectors $\bar{x}$ and $\bar{u}$ are optimal for these problems. $\square$

The analog of conditions (16.3) for the pair $\{(13.4),(13.4^*)\}$ will be the condition

$$\bar{u}^T[A\bar{x} - b] = 0. \tag{16.4}$$

The optimality conditions for the *LP* problem (for example, of (15.1)) can be expressed in terms of *saddle point* $[\bar{x}, \bar{u}]$ for the Lagrange function $L(x, u) = (c, x) - (u, Ax - b)$, which corresponds to problem (15.1). The definition of the saddle point $[\bar{x}, \bar{u}] \geq 0$ for $L(x, u)$ is as follows

$$L(x, \bar{u}) \underset{\forall x \geq 0}{\leq} L(\bar{x}, \bar{u}) \underset{\forall u \geq 0}{\leq} L(\bar{x}, u),$$

which means that

$$(c, x) - (Ax - b, \bar{u}) \underset{\forall x \geq 0}{\leq} (c, \bar{x}) - (A\bar{x} - b, \bar{u}) \underset{\forall u \geq 0}{\leq} (c, \bar{x}) - (A\bar{x} - b, u). \quad (16.5)$$

This relation can be expressed in the following form:

$$(c - A^T \bar{u}, x) + (b, \bar{y}) \underset{\forall x \geq 0}{\leq} (c - A^T \bar{u}, \bar{x}) + (b, \bar{u}) \underset{\forall u \geq 0}{\leq} (c, \bar{x}) - (A\bar{x} - b, u).$$

$$(16.6)$$

Theorem 16.3. *Condition* (16.3) *hold if and only if* $[\bar{x}, \bar{u}]$ *is the saddle point for the Lagrange function* $L(x, u)$.

Proof. We suppose that conditions (16.3) hold. This denotes, in particular, that $\bar{x} \in M$, $\bar{u} \in M^*$ and $(c, \bar{x}) = (b, \bar{u})$. The right inequality in (16.5) is evident since

$$(A\bar{x} - b, \bar{u}) = 0 \quad \text{and} \quad A\bar{x} - b \leq 0.$$

The left inequality will be as follows

$$(c - A^T \bar{u}, x) \leq 0, \qquad \forall x \geq 0$$

if we take into account (16.6). As $c - A^T \bar{u} \leq 0$, then the left inequality holds also.

Conversely, we suppose that $[\bar{x}, \bar{u}]$ is the saddle point for $L(x, u)$. We show now that conditions (16.3) hold. The right inequality in (16.5) we represent in the form

$$(\bar{u}, A\bar{x} - b) \underset{\forall u \geq 0}{\geq} (A\bar{x} - b, u).$$

Now, we prove that $A\bar{x} - b \leq 0$. Really, if $A\bar{x} - b \not\leq 0$, then one coordinate, suppose j' of the vector $A\bar{x} - b$ should be positive. However, if we take j' coordinate of the vector $u \geq 0$ sufficiently large, and all other coordinates as zeros, we should obtain the contradiction with the inequality

$$(\bar{u}, A\bar{x} - b) \geq (A\bar{x} - b, u), \qquad \forall u \geq 0.$$

Further, if $(\bar{u}, A\bar{x} - b) \neq 0$, i. e., $(\bar{u}, A\bar{x} - b) =: -\alpha < 0$, then we should have $0 > -\alpha \geq 0$ for $u = 0$, which is inconsistently also. So, we have established the first relation from (16.3). We shall prove now the second one. The left inequality from (16.5) we rewrite as follows

$$(c - A^T \bar{u}, x) \leq (c - A^T \bar{u}, \bar{x})$$

taking into account (16.6). Similarly, as before, we show that $c - A^T u \leq 0$ and $(c - A^T \bar{u}, \bar{x}) = 0$. $\square$

17. INFORMATIVE INTERPRETATION OF OPTIMALITY CONDITIONS

In Sections 13 and 14 we had given the economic interpretation for problems (15.1) and (15.1*). We shall give now similar interpretation for optimality conditions (16.3). These conditions can be represented as follows:

$$\bar{u}_j \left[\sum_{i=1}^{n} a_{ji}\bar{x}_i - b_j \right] = 0, \qquad j = 1,\ldots,m, \tag{17.1}$$

$$\bar{x}_i \left[\sum_{i=1}^{m} a_{ji}\bar{u}_j - c_i \right] = 0, \qquad i = 1,\ldots,n. \tag{17.2}$$

This conditions are equivalent to those conditions

$$\left.\begin{array}{l} \text{if} \quad \bar{u}_j > 0, \quad \text{then} \quad \sum_{i=1}^{n} a_{ji}\bar{x}_i = b_j, \quad j = 1,\ldots,m, \\[4mm] \text{if} \quad \sum_{i=1}^{n} a_{ji}\bar{x}_i < b_j, \quad \text{then} \quad \bar{u}_j = 0, \quad j = 1,\ldots,m; \end{array}\right\} \tag{17.3}$$

$$\left.\begin{array}{l} \text{if} \quad \bar{x}_i > 0, \quad \text{then} \quad \sum_{j=1}^{m} a_{ji}\bar{u}_j = c_i, \quad i = 1,\ldots,n, \\[4mm] \text{if} \quad \sum_{j=1}^{m} a_{ji}\bar{u}_j > c_i, \quad \text{then} \quad \bar{x}_i = 0, \quad i = 1,\ldots,n. \end{array}\right\} \tag{17.4}$$

Conditions (17.3) may be interpreted as follows: *if the estimate of j resource is positive, then this resource is used completely. Conversely, if the resource is not used completely, then its estimate is equal to zero.*

Conditions (17.4), respectively, may be interpreted so: *if we use the ith technological methods (this means that its intensity is positive), then it is not disadvantaged in the estimates $\bar{u}_1,\ldots,\bar{u}_m$. If the ith technological method is disadvantaged in estimate $\bar{u}_1,\ldots,\bar{u}_m$, then it does not involved in production.*

We show now one useful characteristic of the dual estimates. For this purpose we consider problem (15.1). Assume that the vector of dual estimates $[\bar{u}_1,\ldots,\bar{u}_m]$ for this problem is determined uniquely (i. e., the optimal set for problem (15.1*) has the only vector $\bar{u}$). We consider the optimum for problem (15.1) as the function of the ingredient vector b. Let $\tilde{f}(b)$

be its notation. The function $\tilde{f}(b)$ is determined in a certain neighborhood of the point b (for our assumptions) and has any directional derivative $(s = [s_1, \ldots, s_m])$

$$\frac{\partial \tilde{f}(b)}{\partial s} = \lim_{t \to +0} \frac{\tilde{f}(b + ts) - \tilde{f}(b)}{t} = \sum_{j=1}^{m} \bar{u}_j s_j. \qquad (17.5)$$

Really, taking into account the duality theorem, we have

$$\tilde{f}(b) = (b, \bar{u}); \qquad \tilde{f}(b + ts) = (b + ts, \bar{u})$$

for $t > 0$ sufficiently small, whence required relation (17.5) follows.

In particular, if $s = [0, \ldots, s_j = 1, 0, \ldots, 0]$, then we obtain

$$\frac{\partial \tilde{f}(b)}{\partial b_j} = \bar{u}_j, \qquad j = 1, \ldots, m.$$

These relations denote that the dual estimate $\bar{u}_j$ of jth ingredient plays the role of measuring instrument of its efficiency (in the sense of the criterion (c, x)).

18.   MATRIX PLAYS AND DUALITY

We consider here *the play with two players* which have antagonistic interests. We suppose that each player has a certain number of *strategies* (or *pure strategies*) — this means the ways of behavior. The strategies of the first player we numerate with the indices $i = 1, \ldots, n$; the second player we have the strategies with numbers $j = 1, \ldots, m$. If the first player chooses the ith strategy and the second jth strategy, then the result of the game is defined uniquely and is equal to $E(i, j) = a_{ij}$. This result is interpreted as the charge of the second player and the gain of the first player.

Therefore, the play (the goal of the game is unknown) is given by the matrix

$$A = (a_{ij})_{n,m} = \left\| \begin{array}{cccc} a_{11} & a_{12} & \cdots & a_{1m} \\ a_{21} & a_{22} & \cdots & a_{2m} \\ \cdots & \cdots & \cdots & \cdots \\ a_{n1} & a_{n2} & \cdots & a_{nm} \end{array} \right\|.$$

We introduce now the notion of *mixed strategy* (it is the widening of the notion of pure strategy). We shall call the non-negative vector

$x = [x_1, \ldots, x_n]$, $\sum_{i=1}^n x_i = 1$ by the mixed strategy of the first player. Analogously, the non-negative vector $y = [y_1, \ldots, y_m] \geq 0$, $\sum_{j=1}^m y_j = 1$ will be called by the mixed strategy of the second player.

The pure strategies, in this case, may be considered as

$$x^i = [0, \ldots, x_i = 1, 0, \ldots, 0] \quad \text{and} \quad y^j = [0, \ldots, y_j = 1, 0, \ldots, 0]$$

which are the ith pure strategy of the first player and the jth pure strategy of the second player respectively. The *payoff function* we define as follows

$$E(x, y) = \sum_{i,j} a_{ij} x_i y_j,$$

so that $E(x^i, y^j) = E(i, j) = a_{ij}$.

We set now

$$M = \{x = [x_1, \ldots, x_n] \geq 0 \mid \sum_{i=1}^n x_i = 1\},$$
$$M^* = \{y = [y_1, \ldots, y_m] \geq 0 \mid \sum_{j=1}^m y_j = 1\}.$$

One of the natural settings of the play may be obtained from the following considerations. If the first player chooses the strategy $x \in M$, he guarantees the gain $\min_{y \in M^*} E(x, y)$. Therefore, he ma provide to himself the following gain

$$\max_{x \in M} \min_{y \in M^*} E(x, y) = \tilde{v}. \tag{18.1}$$

The analogous approach applied to the second player gives

$$\min_{y \in M^*} \max_{x \in M} E(x, y) = \tilde{v}^*. \tag{18.1*}$$

The goal of the game, in this case, will be as follows. The players have to provide the gain not less than $\tilde{v}$ and the loss not greater than $\tilde{v}^*$ respectively. The goals of the players formulated as above are not contradictive (we shall show below that $\tilde{v}$ and $\tilde{v}^*$ coincide).

We denote this game by the symbol Γ.

Definition 18.1. The three elements $\{\tilde{x}, \tilde{y}, \tilde{v}\}$, $\tilde{x} \in M$, $\tilde{y} \in M^*$, $\tilde{v} \in \mathbb{R}$ are the solution of the game Γ if the following relations

$$E(\tilde{x}, y^j) \geq \tilde{v}, \quad j = 1, \ldots, m, \qquad E(x^i, \tilde{y}) \leq \tilde{v}, \quad i = 1, \ldots, n$$

hold. The vectors $\tilde{x}$ and $\tilde{y}$ in this case are called optimal strategies (of the 1st and 2nd players respectively); $\tilde{v}$ is called *the value* of the game (or *the price*).

Parallel with relations (18.1) and (18.1*) we consider the equivalent problems

$$\max\{v \mid E(x,y) \geq v, \ \forall y \in M^*\}, \tag{18.2}$$
$$\min\{v^* \mid E(x,y) \leq v^*, \ \forall x \in M\},. \tag{18.2*}$$

We show, for example, that (18.1) is equivalent to (18.2). Denote by $\{\bar{x}, \bar{v}\}$ the solution of problem (18.2). Then we have: $E(\bar{x}, y) \geq \bar{v}, \ \forall y \in M^*$; therefore,

$$\bar{v} \leq \min_{y \in M^*} E(\bar{x}, y) \leq \max_{x \in M} \min_{y \in M^*} E(x, y) = \tilde{v}.$$

From the other hand, there does not exist such $x \in M$ that $E(x,y) > \bar{v}$, $\forall y \in M^*$. All this yields

$$\min_{y \in M^*} E(x, y) \leq \bar{v}, \quad \forall x \in M,$$

therefore,

$$\max_{x \in M} \min_{y \in M^*} E(x, y) \leq \bar{v}.$$

Hence, we have $\bar{v} = \tilde{v}$. If $\bar{y}$ is any value for $y \in M^*$ which realizes the equality $E(x, \bar{y}) = \bar{v}$ for $x = \bar{x}$, then the vector $[\bar{x}, \bar{y}]$ solves problem (18.1). Conversely, if the vector $[\tilde{x}, \tilde{y}, \tilde{v}]$ solves problem (18.1) then $\{\tilde{x}, \tilde{v}\}$ solves problem (18.2).

The equivalence of (18.1*) and (18.2*) is established analogously.

We rewrite (18.2) and (18.2*) in the form of *LP* problems

$$\max\left\{v \mid E(x, y^j) \equiv \sum_{i=1}^{n} a_{ij} x_i \geq v, \right.$$

$$\left. j = 1, \ldots, m, \ \sum_{i=1}^{n} x_i = 1, \ x \geq 0\right\} (= \tilde{v}), \tag{18.3}$$

$$\min\left\{v^* \mid E(x^i, y) \equiv \sum_{j=1}^{m} a_{ij} y_j \leq v^*, \right.$$

$$i = 1, \ldots, m, \ \sum_{j=1}^{m} y_j = 1, \ y \geq 0 \Big\} \ (= \tilde{v}^*). \qquad (18.3^*)$$

Really, if $\bar{x}$ is admissible for (18.2), then it is admissible for (18.3) also. Conversely, if $\bar{x}$ is admissible for (18.3) and $y \in M^*$ is arbitrary, then the following chain holds

$$E(\bar{x}, y^j) \geq v, \quad j = 1, \ldots, m \implies$$
$$\implies y_j E(\bar{x}, y^j) \geq y_j v, \quad j = 1, \ldots, m \implies$$
$$\implies \sum_{j=1}^{m} y_j E(\bar{x}, y^j) = E(\bar{x}, y) \geq \sum_{j=1}^{m} y_j v = v,$$

which means that $\bar{x}$ is admissible for (18.2).

The equivalence of (18.2*) and (18.3*) is proved analogously.

Note that problem (18.3) (problem (18.3*) also) is solvable for any matrix $A = [a_{ij}]_{n,m}$. This follows from boundedness of v (from above) in the set of admissible $[x, v]$:

$$v \leq \sum_{i=1}^{n} a_{ij} x_i \leq \max_{(i)} |a_{ij}| \sum_{i=1}^{n} = \max_{(i)} |a_{ij}|.$$

If we pass from problem (18.3) by the scheme $\overset{(*)}{\to}$ (from Section 14) to the dual problem with the system $u_1, \ldots, u_m$, then changing y_j by $-u_j$ we obtain problem (18.3*). Therefore, by the duality theorem, for the LP problem we may say that the optimal values for (18.3) and (18.3*) coincide; i.e. $\tilde{v} = v^*$. Therefore, the following theorem holds.

Theorem 18.1 [Neyman]. *The values of problem* (18.1) *and* (18.1*) *coincide, namely:*

$$\max_{x \in M} \min_{y \in M^*} E(x, y) = \min_{y \in M^*} \max_{x \in M} E(x, y).$$

The following theorem is the result of the reducibility of Γ to solving the LP problems (18.3) and (18.3*).

Theorem 18.2. *The following statements are equivalent*

1) $\{\tilde{x}, \tilde{y}, \tilde{v}\}$ *is a solution of the game* Γ;

2) $\{\tilde{x}, \tilde{v}\}$ and $\{\tilde{y}, \tilde{v}\}$ are solutions of problems (18.3) and (18.3*) respectively.

Proof. Really, 2) yields 1) by the definition of the game Γ. Further, if $\{\tilde{x}, \tilde{y}, \tilde{v}\}$ is a solution of Γ, then the vector $[\tilde{x}, \tilde{v}]$ is admissible for (18.3) and $[\tilde{y}, \tilde{v}]$ is admissible for (18.3*). In this case, the values v and v^* coincide and are equal to $\tilde{v}$. By the corollary to Theorem 15.1 these vectors are optimal for problems (18.3) and (18.3*) respectively.   □

From this theorem it follows that the set of the vectors $[\tilde{x}, \tilde{y}]$ which realize in (18.1) the equality $E(\tilde{x}, \tilde{y}) = \tilde{v}$ (or, which is the same, $\{\tilde{x}, \tilde{y}, \tilde{v}\}$ is the solution of the game Γ) coincide with $\tilde{M} \times \tilde{M}^* \subset \mathbb{E}_n \times \mathbb{E}_m$. Here $\tilde{M}$ and $\tilde{M}^*$ are optimal sets for problems (18.3) and (18.3*).

19.   THE THEOREM ON MARGINAL VALUES

Duality for the LP problem allows us to investigate certain differential properties of the optimum function $\tilde{f}(y) := \operatorname{opt} L$, where L is an arbitrary LP problem, $y = [A, b, c]$. We suppose that

$$L: \quad \max\{(c, x) \mid Ax \leq b\}. \tag{19.1}$$

The theorem on marginal values for the problem L is the theorem on explicit representation of the directional derivative of the optimum function $\tilde{f}(y)$ with respect to the arbitrary direction l from the space of the vector y. We take the vector $[\Delta A, \Delta b, \Delta c]$ as the vector l. Particular case of this situation is realized by relation (17.5), where l is equal to Δb in the space of vector b (as the information parameter of the problem L), i. e., $l = \Delta b$. Relation (17.5) in this notations will be as follows:

$$\frac{\partial \tilde{f}(y)}{\partial l} = (\bar{u}, \Delta b), \tag{19.2}$$

where $\bar{u} = \arg L^*$. If the optimal vector $\bar{x}$ of the initial problem (19.1) is unique, then the analog of relation (19.2) is as follows

$$\frac{\partial \tilde{f}(y)}{\partial l} = (\bar{x}, \Delta c), \tag{19.3}$$

where $l = \Delta c$. In the general case the formula for $\partial \tilde{f}(y)/\partial l$ can not be obtained in the framework of linear analysis of the problem L.

In this case we shall take a part of the general theorem for the parameter problems of convex programming (see Appendix, Theorem 14).

Theorem 19.1. *Suppose the problem L satisfies the conditions*

1) $\exists p : Ap < b$;

2) the optimal sets $\operatorname{Arg} L^* =: \tilde{M}$ *and* $\operatorname{Arg} L^* =: \tilde{M}^*$ *are nonempty and bounded.*

If $l = [\Delta A, \Delta b, \Delta c]$ *is an arbitrary direction in the space of the vector* $y = [A, b, c]$ *then*

$$\frac{\partial \tilde{f}(y)}{\partial l} = \max_{\bar{x} \in \tilde{M}} \min_{\bar{u} \in \tilde{M}^*} [(\Delta c, \bar{x}) - (\Delta A \bar{x} - \Delta b, \bar{u})]. \qquad (19.4)$$

Therefore, if L *and* L^* *are solvable in the unique points* $\bar{x}$ *and* $\bar{u}$, *then*

$$\frac{\partial \tilde{f}(y)}{\partial l} = (\Delta c, \bar{x}) - (\Delta A \bar{x} - \Delta b, \bar{u}). \qquad (19.5)$$

Corollary 19.1. *In assumptions which give (19.5) the following relations hold*

$$\frac{\partial \tilde{f}(y)}{\partial b_j} = \bar{u}_j, \quad \frac{\partial \tilde{f}(y)}{\partial c_i} = \bar{x}_i, \quad \frac{\partial \tilde{f}(y)}{\partial a_{ji}} = -\bar{u}_j \bar{x}_i. \qquad (19.6)$$

Here $\bar{u}^T = [\bar{u}_1, \dots, \bar{u}_m]$, $\bar{x}^T = [\bar{x}_1, \dots, \bar{x}_n]$.

In the notations of the optimum function $\tilde{f}(y)$ we shall use those part of the informative vector which we are interested: $\tilde{f}(b)$, $\tilde{f}(b_j)$ and so on.

The following remark will be connected with informative representation of dual estimates of $\bar{u}_j$, $j = 1, \dots, m$. When the optimal vector $\bar{u}$ of the problem L^* is unique, there exists a neighbourhood $v(b)$ of the vector $b \in V^0(b)$ (V^0 is the interior of the set V) such that

$$\bar{b} \in V(b) \Longrightarrow \operatorname{Arg} \min_{u \in M^*} (\bar{b}, u) = \{\bar{u}\}.$$

Then for a certain small $t_0 > 0$ and $\forall t \in (0, t_0]$ we have

$$\frac{\tilde{f}(b + t\Delta b) - \tilde{f}(b)}{t} = \frac{(b + t\Delta b, \bar{u}) - (b, \bar{u})}{t} = (\Delta b, \bar{u}),$$

i. e.,

$$\tilde{f}(b + t\Delta b) = \tilde{f}(b) + t(\Delta b, \bar{u}). \qquad (19.7)$$

If $t_0 > 1$, which means that we may take $t = 1$ providing (19.7), then, for $\Delta b = [0, \ldots, \Delta b_j = 1, 0, \ldots, 0]^T$ from (19.7) we obtain

$$\tilde{f}(b_j + 1) = \tilde{f}(b_j) + \bar{u}_j, \qquad j = 1, \ldots, m.$$

These relations show that the estimates of $\bar{u}_j$ measure the efficiency of the unity of jth resource b_j in the economic interpretation of the problem L from Section 12. This clarifies the interpretations of relation (17.5).

In addition we shall introduce the explicit form of the direction derivative of the payoff function. for the matrix play in the mixed strategies (see Section 18). We suppose that

$$\max_{x \in M} \min_{y \in M^*} x^T A y =: E(A).$$

where $M = \{x \geq 0 \mid \sum_{i=1}^n x_i = 1\}$, $M^* = \{y \geq 0 \mid \sum_{j=1}^m y_j = 1\}$. If $l = \Delta A$, then

$$\frac{\partial E(A)}{\partial l} = \max_{\bar{x} \in \tilde{M}} \min_{\bar{u} \in \tilde{M}^*} \bar{x}^T \Delta A \bar{y}, \qquad (19.8)$$

where $\tilde{M}$ and $\tilde{M}^*$ are the sets of optimal strategies of the first and second players respectively.

The derivation of relation (19.8) is base on relation (19.4) and the reduction of the matrix game to the pair of mutually dual problems of linear programming (18.3) and (18.3*).

20. THE METHOD OF EXACT PENALTY FUNCTIONS IN LINEAR PROGRAMMING

The methods of penalty functions are meant the ways of equivalent (or asymptotically equivalent) reducing the LP problem to the new optimization problem. In this new problem a certain part of restrictions are taking into account in the reconstructed goal function. Such reducing has a lot of realizations and we consider only one of them.

We suppose that LP problem has the form

$$L: \quad \max\{(c, x) \mid Ax \leq b, \ x \in M_0\}, \qquad (20.1)$$

where M_0 is the rectangle given by a certain system of linear inequalities. The cases $M_0 = \mathbb{E}_n$ and $M_0 = \mathbb{R}_+^n = \{x \geq 0\}$ are possible. In whole, the reconstructed goal function has the form $\Phi(R, x) = (c, x) - \varphi(R, x)$, where the function $\varphi(R, x)$ is the penalty function for violation of the inclusion $x \in \{x \mid Ax \leq b\}$, R is the parameter vector scaling the penalty measure and determining the the measure of proximity of solutions of L and of the reconstructed problem

$$\max\{\Phi(R, x) \mid x \in M_0\}. \tag{20.2}$$

The function $\varphi(R, x)$ which provides the equivalence of problems (20.1) and (20.2) for certain values of R is called the exact penalty function.

We shall give now one of the important constructions of the exact penalty function. Divide the system $Ax \leq b$ in arbitrary way on m_0 subsystems $A_j x \leq b^j$, $j = 1, \ldots, m_0$, so that

$$A = \begin{bmatrix} A_1 \\ \vdots \\ A_{m_0} \end{bmatrix}, \qquad b = \begin{bmatrix} b_1 \\ \vdots \\ b^{m_0} \end{bmatrix}.$$

Let $\|\cdot\|_j$ be an arbitrary system of norms of spaces which have similar dimension as the vector b_j, $j = 1, \ldots, m_0$. Denote by $\|\cdot\|_j^*$ the norm conjugate with the norm $\|\cdot\|_j$, which is defined as follows

$$\|y\|_j^* = \sup_{\|x\|_j \leq 1} (x, y).$$

The problem will be formulated so

$$P: \quad \max\left\{ \Phi(R, x) := (c, x) - \sum_{j=1}^{m_0} R_j \|(A_j x - b^j)^+\|_j \mid x \in M_0 \right\}, \tag{20.3}$$

where $\{R_j\}_1^{m_0}$ are positive parameters. In this case, the function $\varphi(R, x) = \sum_{j=1}^{m_0} R_j \|(A_j x - b^j)^+\|_j$ is the penalty function. For R_j sufficiently large, the problems L and P will be equivalent in the sense opt L = opt P and Arg L = Arg P. This means their optimal sets and optimal values coincide. In order to make the formulation more precise we write the problem dual to L. Suppose $M_0 = \{x \geq 0 \mid Bx \leq d\}$. The following problem

$$L^*: \quad \min\{(b, u) + (d, v) \mid A^T u + B^T v \geq c, \ [u, v] \geq 0\} \tag{20.1*}$$

will be dual to L. Set $[\tilde{u}, \tilde{v}] = \arg(20.1^*)$, $\tilde{u}^T = [\tilde{u}_1, \ldots, \tilde{u}_{m_0}]$ is the partition of $\tilde{u}$ into the fragments correspondent to partition of A onto the submatrices A_j, $j = 1, \ldots, m_0$.

Theorem 20.1. *Let the problem L (i.e., (20.1)) be solvable. If $R_j \geq \|\tilde{u}_j\|_j^*$, $j = 1, \ldots, m_0$ then* opt $L =$ opt P *and* Arg $L \subset$ Arg P; *if $R_j > \|\tilde{u}_j\|_j^*$, $j = 1, \ldots, m_0$, then* Arg $L =$ Arg P.

Proof. The following two relations

$$(\tilde{u}_j, (A_j\bar{x} - b^j)^+) \leq \|\tilde{u}_j\|_j^* \|(A_j\bar{x} - b^j)^+\|_j,$$
$$(c, \tilde{x}) = (b, \tilde{u}) + (d, \tilde{v}),$$

we shall use in the below proof. Here $\tilde{x} \in$ Arg L, $\bar{x} \in M$. The first of these relations follows from the definition of the conjugate norm, the second relation expresses the duality theorem.

As $\Phi(R, \tilde{x}) = (c, \tilde{x}) =$ opt L then opt $P \geq$ opt L. Now, we shall prove the inverse inequality. For each $\bar{x} \in M_0$, we have

$$\Phi(R, \tilde{x}) = (c, \bar{x}) - \sum_{j=1}^{m_0} R_j \|(A_j\bar{x} - b^j)^+\|_j$$

$$\leq (A^T\tilde{u} + B^T\tilde{v}, \bar{x}) - \varphi(R, \bar{x})$$

$$= (A\bar{x} - b, \tilde{u}) + (B\bar{x} - d, \tilde{v}) + (b, \tilde{u}) + (d, \tilde{v}) - \varphi(R, \bar{x})$$

$$= \sum_{j=1}^{m_0} (A_j\bar{x} - b^j, \tilde{u}) + (B\bar{x} - d, \tilde{v}) + (c, \tilde{x}) - \varphi(R, \bar{x})$$

$$\leq \sum_{j=1}^{m_0} \|(A_j\bar{x} - b^j)^+\|_j \|\tilde{u}_j\|_j^* + \text{opt } L - \varphi(R, \bar{x})$$

$$= \text{opt } L - \sum_{j=1}^{m_0} (R_j - \|\tilde{u}_j\|_j^*) \|(A_j\bar{x} - b^j)^+\|_j \leq \text{opt } L. \quad (20.4)$$

Hence it follows that opt $P \leq$ opt L; therefore, opt $L =$ opt P (for $R_j \geq \|\tilde{u}_j\|_j^*$, $j = 1, \ldots, m_0$). As $\Phi(R, \bar{x}) =$ opt P, then $\bar{x} \in$ Arg P; therefore, Arg $L \subset$ Arg P.

Suppose $\bar{x} \in$ Arg P and $R_j \geq \|\tilde{u}_j\|_j^*$, $j = 1, \ldots, m_0$. Inequality (20.4) yields

$$\text{opt } P \leq \text{opt } P - \sum_{j=1}^{m_0} (R_j - \|\tilde{u}_j\|_j^*) \|(A_j\bar{x} - b^j)^+\|_j \leq \text{opt } P,$$

therefore, all the terms in the sum are equal to zero. Hence it follows that $(A_j\bar{x} - b^j)^+ = 0$, i.e., $A_j\bar{x} - b^j \leq 0$, $j = 1, \ldots, m_0$. So the vector $\bar{x} \in$ Arg P

satisfies the property of being admissible for the problem L which, together with opt L = opt P yields $\bar{x} \in \text{Arg } L$. Therefore, $\text{Arg } P \subset \text{Arg } L$, and, therefore, $\text{Arg } L = \text{Arg } P$. □

We shall represent now the particular cases of the function $\varphi(R, x)$ in problem (20.3) which provide Theorem 20.1. These cases are obtained as a result of partition of $Ax \leq b$ onto the subsystems $A_j x \leq b^j$, $j = 1, \ldots, m_0$ and as a result of choosing of arbitrary norms $\{\|\cdot\|\}_1^{m_0}$:

$$\varphi_1(R, x) = (R, (Ax - b)^+);$$

$$\varphi_2(R_0, x) = R_0 \|(Ax - b)^+\|, \text{ where } \|\cdot\| \text{ is an arbitrary norm;}$$

$$\varphi_3(R_0, x) = R_0 \max_{(j)} l_j^+(x), \text{ where } l_j(x) \text{ are the left parts of the system of inequalities } Ax - b \leq 0, R_0 > 0.$$

Problem (20.3), generally speaking, is not linear. However, it is the problem of convex programming. The theory of such problems is well developed (see Appendix) and the methods for solving these problems are numerous. If the norms $\{\|\cdot\|\}_1^{m_0}$ are piecewise-linear, for example, $\sum_{i=1}^s |z_i|$, $\max_{(i)} |z_i|$ for $z \in \mathbb{R}^s$, then (20.3) will be the problem of piecewise-linear convex programming. Such problem can be transformed to the *LP* problem. Suppose, for example, $\varphi(R_0, x) = R_0 \max_{(j)} l_j^+(x)$, then the problem

$$\max\{(c, x) - R_0 \max_{(j)} l_j^+(x) \mid x \in M_0\}$$

can be rewritten in the form

$$\max\{(c, x) - R_0 t \mid l_j(x) \leq t, \ j = 1, \ldots, m, \ t \geq 0, \ x \in M_0\};$$

and it is the *LP* problem with the variable vector $[x, t]$.

Note, in addition, the the function $\varphi(R, x)$ (in the above method) is not differentiable (generally speaking). This is so because the vector functions $(A_j x - b^j)^+$ are piecewise-linear and the norm $\|\cdot\|$ are not differentiable. In spite of the fact (the property to be differentiable is of great importance in the gradient methods) the difficulties can be overcome by introducing special penalty functions. These functions, however, provide only asymptotic equivalence of initial and reduced problems.

First, we consider a particular case of the penalty function $\varphi(R, x)$ which modifies the function $\varphi_1(R, x) = (R, (Ax - b)^+) = \sum_{j=1}^m R_j l_j^+(x)$, namely $\varphi(R, x) = \sum_{j=1}^m R_j l_j^{+2}(x)$. This function, unlike $\varphi_1(R, x)$, is differentiable (with the gradient $\nabla\varphi(R, x) = 2\sum_{j=1}^m R_j l_j^+(x) a_j$). For this problem the equivalent reduction is impossible, but the following statement holds.

Theorem 20.2. *Suppose that in* (20.2)

$$\Phi(R, x) = (c, x) - \sum_{j=1}^{m_0} R_j \|(A_j x - b^j)^+\|_j^2. \tag{20.5}$$

Then, for arbitrary $R > 0$, the estimate

$$\text{opt } L \le \text{opt } (20.2) \le \text{opt } L + \sum_{j=1}^{m_0} \frac{(\|\tilde{u}_j\|_j^*)^2}{4R_j} \tag{20.6}$$

holds.

Proof. The left inequality in (20.6), as in Theorem 20.1, is evident. For $x \in M_0$, taking into account the calculations in (20.4), we obtain

$$\Phi(R, x) \le \text{opt } L + \sum_{j=1}^{m_0} \|\tilde{u}_j\|_j^* \|(A_j x - b^j)^+\|_j - \sum_{j=1}^{m_0} R_j \|(A_j x - b^j)^+\|_j^2$$

$$= \text{opt } L + \sum_{j=1}^{m_0} \frac{(\|\tilde{u}_j\|_j^*)^2}{4R_j} - \sum_{j=1}^{m_0} \left(\sqrt{R_j} \|(A_j x - b^j)^+\|_j - \frac{\|\tilde{u}_j\|_j^*}{2\sqrt{R_j}} \right)^2$$

$$\le \text{opt } L + \sum_{j=1}^{m_0} \frac{(\|\tilde{u}_j\|_j^*)^2}{4R_j}.$$

Therefore, the right inequality in (20.6) holds also. $\quad\square$

Corollary 20.1. *In the conditions of Theorem 20.2, we have*

$$\text{opt } (20.2) \to \text{opt } L \quad \text{for} \quad \max_{(j)} R_j \to +\infty.$$

In this case we say that problems (20.1) *and* (20.2) *are asymptotically equivalent if the function $\Phi(R, x)$ is considered in form* (20.5).

21. *LP* **PROBLEMS WITH SEVERAL CRITERION FUNCTIONS**

The admissible vector of *LP* problem, interpreted as a plan, may be characterized by many criterions. In the informative sense such situation is even more typical than the case with one criterion function. Suppose these criterion functions are $f_j(x) = c_j^T x$, $j = 1, \ldots, k$, and the system of restrictions is as follows $Ax \leq b$, $x \geq 0$. Even if each of the criterion functions must attain the maximal value, the optimization model will not be uniquely determined. So, we must clarify the setting. Below we shall consider the two classical settings.

21.1. The Pareto optimization model

We set $[c_1^T x, \ldots, c_k^T x]^T = C^T x =: F(x)$. Here $C = [c_1, \ldots, c_k]$, c_j are column vectors.

The vector $\bar{x} \in M = \{x \geq 0 \mid Ax \leq b\}$ we call *Pareto maximal* (or *π-maximal*) with respect to the system of criterion functions $F(x)$ if the following implication holds

$$F(x) \geq F(\bar{x}), \qquad z \in M \;\Rightarrow\; F(z) = F(\bar{x}). \tag{21.1}$$

This denotes that if $\bar{x}$ is a π-maximal point, then we cannot pass to another point improving one of the criterion without worsening at least one of the others.

The following definition is one of equivalent identifications of Pareto optimal point $\bar{x}$:

The point $\bar{x}$ is π-maximal if and only if the vector $[\bar{x}, \underbrace{0, \ldots, 0}_{k}]^T$ is optimal for the LP problem

$$\max\left\{ \sum_{j=1}^{k} y_j \;\middle|\; Ax \leq b,\; C^T x \geq C^T \bar{x} + y,\; x \geq 0,\; y \geq 0 \right\}. \tag{21.2}$$

This characterization of the π-optimality we shall use below.

The Pareto optimal problem we shall write in the form

$$\max_{\pi}\{ C^T x \mid Ax \leq b,\; x \geq 0 \}. \tag{21.3}$$

The sense of this problem is theoretic and algorithmic identification of the set Arg (21.3) of π-maximal points (or of its part), or, simple, to find an element from Arg (21.3).

In many-criterion optimization we apply the general approach, when we reduce initial problem to the problem with one criterion $f(f_1(x), \ldots, f_k(x))$. This criterion is constructed with the use of the functions $\{f_j(x)\}_1^k$. The important moment in this reducing is the linear convolution

$$f_R(x) := \sum_{j=1}^{k} R_j f_j(x), \qquad R_j > 0, \quad j = 1, \ldots, k.$$

We consider now the problem:

$$\max\left\{ \sum_{j=1}^{k} R_j c_j^T x = (C^T x, R) \mid Ax \le b, \ x \ge 0\right\}, \tag{21.4}$$

where $R = [R_1, \ldots, R_k]^T$. Passing to problem (21.4) we call *scalarization* of problem (21.3). From the definition of π-optimal point it follows that

$$\mathrm{Arg}\,(21.4) \subset \mathrm{Arg}\,(21.3). \tag{21.5}$$

Really, all the points from $\mathrm{Arg}\,(21.3)$ can be obtained from problems (21.4) for various collections of the parameter $R > 0$.

For R fixed, problem (21.4) is the LP problem; the problem

$$\min\{(b, u) \mid A^T u \ge Ct, \ u \ge 0\} \tag{21.4*}$$

is dual to this problem.

The optimal property of $\bar{x} \in M$ for (21.4) is as follows: there exists vector $\bar{u} \ge 0$, $A^T \bar{u} \ge CR$ such that $(b, \bar{u}) = (CR, \bar{x})$ (see Section 16). We write here all the relations which define the optimal property of $\bar{x}$ for (21.4):

$$A\bar{x} \le b, \ \bar{x} \ge 0, \quad A^T \bar{u} \ge CR, \ \bar{u} \ge 0, \quad (b, \bar{u}) = (CR, \bar{x}). \tag{21.6}$$

We shall use these relations when proving the basic theorem.

Theorem 21.1. *1. If the vector $\bar{x}$ is optimal for (21.4), then $\bar{x}$ is π-optimal for (21.3), i. e., inclusion (21.5) holds.*
2. If $\bar{x}$ is π-optimal for (21.3), then for a certain $R > 0$ we have $\bar{x} \in \mathrm{Arg}\,(21.4)$.

Proof. We had noted above that inclusion (21.5) holds. We prove now the second part of the theorem. We have to choose such $R > 0$ that for $\bar{x} \in \mathrm{Arg}\,(21.3)$ the optimality conditions (21.6) hold. The dual problem (21.2) will be as follows

$$\min\{(b, u) - (C^T \bar{x}, v) \mid A^T u - Cv \ge 0, \ v \ge E, \ u \ge 0, \ v \ge 0\}. \tag{21.2*}$$

Here u is the vector of dual variables correspondent to the system of restrictions $Ax \le b$; v is the vector correspondent to the system of restrictions $C^T x - y \ge C^T \bar{x}$, $E = \underbrace{[1, \ldots, 1]}_{k}^T$. The optimal property of the vector

$[\bar{x}^T, 0, \ldots, 0]$ for (21.2) will be written in the form: *there exist $\bar{u} \geq 0$, $\bar{v} \geq 0$*
$\underbrace{\qquad\qquad}_{k}$
such that

$$A^T \bar{u} - C\bar{v}, \qquad \bar{v} \geq E\ (> 0), \qquad 0 = (b, \bar{u}) - (C^T \bar{x}, \bar{v}). \qquad (21.7)$$

Here 0 in the equality is the zero value of optimum in problem (21.2). We set $\bar{v} = R$ and rewrite relation (21.7)

$$A^T \bar{u} \geq CR, \qquad \bar{u} \geq 0, \qquad (b, \bar{u}) = (CR, \bar{x}), \qquad R > 0.$$

They coincide with the optimality condition (21.6) of the vector $\bar{x}$ for problem (21.4). So, we have chosen $R > 0$ such that $\bar{x} \in \mathrm{Arg}\,(21.4)$. $\quad\square$

Corollary 21.1. *The following relation holds*

$$\mathrm{Arg}\,(21.3) = \bigcup_{R>0} \mathrm{Arg}\,(21.4). \qquad (21.8)$$

Remark 21.1. The proof of theorem is constructive: it gives the way of choosing the convolution parameter R which provides the inclusion $\bar{x} \in \mathrm{Arg}\,(21.4)$. This parameter is derived from the linear (relatively u and R) system

$$A^T u \geq CR, \qquad (C^T \bar{x}, R) = (b, u), \qquad R > 0. \qquad (21.9)$$

As the system is homogeneous relatively $[u, R]$, we can replace the rigorous inequality $R > 0$ in (21.9) by $R \geq E$, i.e., $R_j \geq 1$, $j = 1, \ldots, k$ (for example). In this connection, the following theorems will be useful.

Theorem 21.2. *The vector $\bar{x}$ admissible for problem (21.3) will be π-optimal if and only if the system (21.9) is compatible relatively u and R.*

Theorem 21.3. *Problem (21.3) is solvable if and only if each of the linear systems $Ax \leq b$, $x \geq 0$ and $A^T u \geq CR$, $u \geq 0$ is solvable.*

Proof. Really, if (21.3) is solvable, then (21.4) is solvable for a certain $R > 0$. Therefore, (21.4*) is solvable also, which proves necessity. Conversely, solvability of systems yields solvability of (21.4*); therefore, problem (21.3) is solvable also. $\quad\square$

Theorem 21.4. *There exists a finite number of vectors $R^t > 0$ which provide equality (21.8), i. e.,*

$$\text{Arg}\,(21.3) = \bigcup_{t=1}^{N} \text{Arg}\,(21.4)\big|_{R=R^t}. \qquad (21.10)$$

Proof. Really, $\text{Arg}\,(21.4)$, for a certain $R > 0$, is a k-face of the polyhedron $M = \{x \geq 0 \mid Ax \leq b\}$ (Theorem 7.2). As the polyhedron M has a finite number of k-faces, then, taking one R^t from each k-face, we obtain representation (21.10). □

21.2. The model of lexicographic (successive) optimization

Another way of formalization of the problem with several criteria is ordering these criteria (by the importance) and organizing subsequent optimization. We suppose that $p = (k, \ldots, 1)$ is the order of indices of the functions $\{c_t^T x\}_1^k$; so, $c_k^T x$ is the most important criterion, then $c_{k-1}^T x$ and so on. We construct the sequence of problems

$$\max\{c_k^T x \mid Ax \leq b,\ x \geq 0\}, \qquad (21.11_k)$$
$$\max\{c_{k-1}^T x \mid x \in \text{Arg}\,(21.11_k)\}, \qquad (21.11_{k-1})$$
$$\cdots\cdots\cdots\cdots\cdots\cdots\cdots\cdots\cdots,$$

$$\boxed{\max\{c_1^T x \mid x \in \text{Arg}\,(21.11_2)\}.} \qquad (21.11_1)$$

The sense of this subsequent optimization by the system of the criterion functions $\{c_t^T x\}_1^k$ consists in its final problem (21.11_1). We shall write this problem briefly

$$\max_{p}\{C^T x \mid Ax \leq b,\ x \geq 0\}, \qquad (21.12)$$

setting $\text{Arg}\,(21.12) = \text{Arg}\,(21.11_1)$.

As in Pareto case, problem (21.12) can be reduced to scalar problem (21.4) even in more strong case. Namely, for certain $R_j,\ j = 1, \ldots, k$, optimal sets of problems (21.12) and (21.4) will coincide.

We consider first the two-step problem of linear optimization:

$$\max\{c^T x \mid Ax \leq b,\ x \geq 0\} \qquad (21.13)$$

is the initial problem. The following problem

$$\max\{c_0^T x \mid x \in \text{Arg}\,(21.13)\} \qquad (21.14)$$

is the final problem. The scalarized problem

$$\max\{c_0^T x + R c^T x \mid Ax \le b, \ x \ge 0\} \tag{21.15}$$

will correspondent to the final problem.

Problem (21.14) can be rewritten as follows:

$$\max\{c_0^T x \mid Ax \le b, \ x \ge 0, \ c^T x \ge \alpha\}, \tag{21.16}$$

where $\alpha := \text{opt}\,(21.13)$.

Lemma 21.1. *Let problem* (21.14) *be solvable (i.e. problem* (21.16) *is solvable). If $\bar{u}_0 \ge 0$ is the dual estimate of the inequality $-c^T x \le -\alpha$ in problem* (21.16) *and $R > \bar{u}_0$, then*

$$\text{Arg}\,(21.4) = \text{Arg}\,(21.15). \tag{21.17}$$

Proof. By Theorem 20.1 on the method of exact penalty functions, problem (21.16), i.e., (21.14), is equivalent (the optimal sets coincide) to the problem

$$\max\{c_0^T x - R(-c^T x + \alpha)^+ \mid x \in M\}, \tag{21.18}$$

where $M := \{x \ge 0 \mid Ax \le b\}$. As $c^T x \le \alpha$, $\forall x \in M$, then the cut-off function "+" in (21.18) may be omitted. Therefore, omitting in (21.18) the term $-\alpha R$, we see that problems (21.14) and (21.15) are equivalent. The lemma is proved. $\square$

Theorem 21.5. *If* (21.12) *is the solvable problem, then there exists a non-empty domain of values of the parameter $R^T = [R_1, \ldots, R_k] > 0$ such that*

$$\text{Arg}\,(21.12) = \text{Arg}\,(21.4). \tag{21.19}$$

Proof. As in the proof of Lemma 21.1, we choose the number $\bar{R}_k > 0$ which provides the equality

$$\text{Arg}\max_{x \in M}\{c_{k-1}^T x + \bar{R}_k c_k^T x\} = \text{Arg}\,(21.11_{k-1}).$$

Then we take $\bar{R}_{k-1}$ so that

$$\text{Arg}\max_{x \in M}\{c_{k-2}^T x + \bar{R}_{k-1}(c_{k-1}^T x + \bar{R}_k c_k^T x)\} = \text{Arg}\,(21.11_{k-2}),$$

and so on. Finally, the parameter $\bar{R}_2 > 0$ provides the final relation

$$\text{Arg}\max_{x \in M}\{c_1^T x + \bar{R}_2(c_2^T x + \bar{R}_3(c_3^T x + \cdots + \bar{R}_k c_k^T x)\ldots)\}$$
$$= \text{Arg}[(21.11_1) \equiv (21.12)]. \quad (21.20)$$

If we set now in (21.20) $R_1 = 1$, $R_2 = \bar{R}_2$, $R_3 = \bar{R}_2\bar{R}_3, \ldots, R_k = \bar{R}_2 \cdots \bar{R}_k$, we obtain the required relation (21.19). □

Remark 21.2. From the proof of Lemma 21.1 it follows that the assumptions of penalty constant R which, by the lemma conditions must be greater than $\bar{u}_0$ (the dual estimate of inequality $c^T x \geq \alpha$ in problem (21.16)) can be done independently of the vector b (the right-hand side of the system of restrictions of problem (21.13). We shall clarify now this fact. Write the problem dual to (21.16)

$$\min\{(b, u) - \alpha u_0 \mid A^T u - u_0 c \geq c_0, \ u \geq 0, \ u_0 \geq 0\}, \quad (21.21)$$

where $\alpha = \text{opt}\,(21.13)$, $[u, u_0]^T$ is a variable vector. We suppose that N is a polyhedron of restrictions of the system, $N \subset \mathbb{E}_{m+1}$. As the restriction system has the rank $m + 1$, then the polyhedron has the vertices which are its $(m + 1)$-faces (Theorem 2.1). Let $P_1, \ldots, P_s$ be these vertices and $u_0^1, \ldots, u_0^s$ be their last coordinates respectively. We set $\tilde{u}_0 = \max_{(t)} u_0^t$. If we take $R > \tilde{u}_0$, then the requirement $R > \bar{u}_0$ in Lemma 21.1 will it a fortiori hold independently of b and α. This means that it is independent of the vertex of the polyhedron N, where the minimum in problem (21.21) be realized.

Hence it follows that in Theorem 21.5 the choice of the parameter vector $R \in \mathbb{E}_k$ can be done independently of the right-hand side of the restriction system in (21.12) also.

This property is not important in this section; however, further, we shall use it essentially. In this case, we shall formulate Theorem 21.5 as follows:

Theorem 21.6. *Let problem* (21.12) *be solvable. Then there exists a non-empty domain of the parameters* $R \subset \mathbb{E}_k$ *determined constructively independently of* b *which provide equality* (21.19).

In Theorem 21.5 problem (21.12) must be solvable. The conditions of its solvability are formulated easily on the basis of solvability of the *LP* problem in the form $M \neq \varnothing \ \& \ M^* \neq \varnothing$, where M and M^* are admissible sets of the direct and dual problems. We suppose, for example, that

$$L: \quad \max\{(c, x) \mid Ax \leq b\},$$

then

$$L^*: \quad \min\{(b, u) \mid A^T u = c, \ u \geq 0\},$$

i.e. $M = \{x \mid Ax \leq b\}$, $M^* = \{u \geq 0 \mid A^T u = c\}$. The condition $M^* \neq \varnothing$ can be rewritten as follows:

$$c \in \text{cone}\{a_j\}_1^m.$$

The modification of this condition is the following condition of solvability of problem (21.12):

$$\left.\begin{array}{l} c_k \in \text{cone}\{a_1^T, \dots, a_m^T, -e_1, \dots, -e_n\}, \\ c_{k-1} \in \text{cone}\{-c_k, J\}, \\ \quad\dots\dots\dots\dots \\ c_1 \in \text{cone}\{-c_2, \dots, -c_k, J\}, \end{array}\right\} \qquad (21.22)$$

where $e_i = [0, \dots, \underset{i}{1}, 0, \dots, 0] \in \mathbb{E}_n$, $J = \{a_1^T, \dots, a_m^T, -e_1, \dots, -e_n\}$. In other words, the following theorem holds.

Theorem 21.7. *The problem (21.12) is solvable if and only if $M = \{x \geq 0 \mid Ax \leq b\} \neq \varnothing$ and conditions (21.22) hold.*

Chapter 3.

Inconsistent problems of linear programming

In this chapter we shall consider the LP problems which are not solvable. In particular, these are the problems whose restriction systems are incompatible (inconsistent). Such LP problems are called *improper problems.*

22. CLASSIFICATION OF IMPROPER PROBLEMS OF LINEAR PROGRAMMING (IP LP)

The initial LP problem we write as follows:

$$L: \ \max\{(c, x) \mid Ax \le b, \ x \ge 0\}, \tag{22.1}$$

and the dual problem to L will be the following problem

$$L^*: \ \min\{(b, u) \mid A^T u \ge c, \ u \ge 0\}. \tag{22.2}$$

As before, the admissible sets of these problems we denote as M and M^* respectively. As we had noted in Section 15 (Theorem 15.4), L or L^* are solvable if and only if $M \ne \varnothing$ or $M^* \ne \varnothing$ respectively. This means that the restriction systems of L and L^* must be solvable.

This moment allows us to connect the incompatibility of the problem L with incompatibility of the system

$$S: \ Ax \le b, \quad x \ge 0; \quad A^T u \ge c, \quad u \ge 0; \quad (c, x) \ge (b, u),$$

which defines the symmetric problem 5 (see Section 15). The connection between the problems L, L^*, and S is given by Theorem 15.5. The incompatibility of the system S is equivalent to incompatibility of its subsystem

$$Ax \leq b, \quad x \geq 0; \qquad A^T u \geq c, \quad u \geq 0,$$

i.e., of at least one of the subsystems $Ax \leq b$, $x \geq 0$ or $A^T u \geq c$, $u \geq 0$ (this follows from the dual Theorem 15.2). Therefore, the classification of $IPLP$ can be done taking into account emptiness ot non-emptiness of the admissible sets M and M^* of the problems L and L^*. The following three alternatives are possible

1) $M = \varnothing$, $M^* \neq \varnothing$;

2) $M \neq \varnothing$, $M^* = \varnothing$;

3) $M = \varnothing$, $M^* = \varnothing$.

According each of these alternatives we call the problem L *improper* of the *first, second,* and *third order* respectively. The case $M \neq \varnothing$ and $M^* \neq \varnothing$ is corresponding to solvability of problems L and L^*.

We shall give the characteristic of each of these three alternatives.

The first alternative is equivalent to the implication

$$M_{\Delta b} := \{x \geq 0 \mid Ax \leq b + \Delta b\} \neq 0 \implies$$
$$\implies L_{\Delta b}: \max_{x \in M_{\Delta b}} (c, x) \quad \text{is solvable.}$$

Really, we suppose that $M^* \neq \varnothing$ and $M_{\Delta b} \neq \varnothing$, then in the pair of mutually dual problems

$$L_{\Delta b}: \max\{(c, x) \mid Ax \leq b + \Delta b, \, x \geq 0\},$$
$$L^*_{\Delta b}: \min\{(b + \Delta b, u) \mid A^T u \geq c, \, u \geq 0\}$$

the asmissible sets are not empty; therefore, $L_{\Delta b}$ is solvable. Conversely, if the implication does not hold, the solvability of $L_{\Delta b}$ yields solvability $L^*_{\Delta b}$; therefore, $M^* \neq \varnothing$.

The improper property of the second order ($M \neq \varnothing$, $M^* = \varnothing$) denotes that the optimal value of the problem L is equal to $+\infty$, i.e.

$$\sup_{x \in M} (c, x) = +\infty.$$

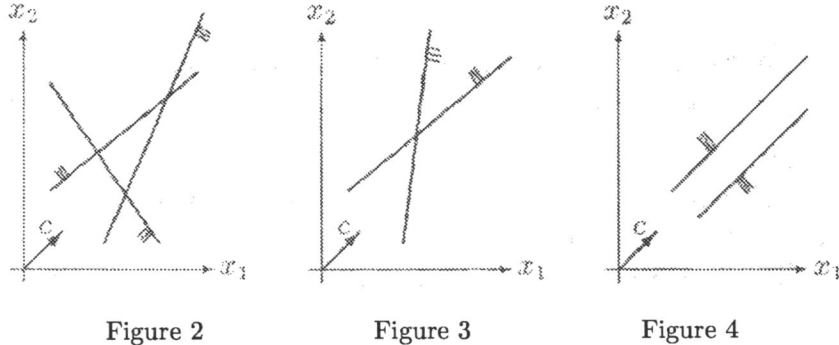

| Figure 2 | Figure 3 | Figure 4 |

Finally, the improper property of the third order, i.e. $M = \varnothing$ and $M^* = \varnothing$ denotes that

$$M_{\Delta b} \neq \varnothing \quad \Longrightarrow \quad \sup_{x \in M_{\Delta b}} (c, x) = +\infty.$$

Therefore, in the case of improper problem of the third order we cannot obtain solvability of the problem L only correcting the vector b in order to provide compatibility of system of restrictions

$$Ax \leq b + \Delta b, \qquad x \geq 0.$$

In this case we must correct the vector c also.

From the property of mutual duality for the pair L and L^* it follows that

1) If L is $IPLP$ of the first order, then L^* is $IPLP$ of the second order.

2) If L is $IPLP$ of the second order, then L^* is $IPLP$ of the first order.

3) If L is $IPLP$ of the third order, then L^* is $IPLP$ of the third order also.

The Figures 2-4 illustrate the cases of improperness of first, second, and third orders respectively

Modelling of practical problems generates, in most cases, the improper problems of the first order.

The following sufficient condition for the problem to be of the first order is tupical for practical problems; there is: for the set of normal vectors of all inequalities-restrictions to be multifold. This means that if the restrictions are as follows:

$$(a_j, x) \leq b_j, \qquad j = 1, \ldots, m, \qquad x \geq 0$$

(the condition $x \geq 0$ is absent), then the system of vectors $\{a_j\}_1^m$ is multifold. The property to be multifold is equivalent to boundedness of the solution polyhedron of the system if it is not empty (Theorem 3.2). Therefore, if the system of normals $\{a_j\}_1^m$ is multifold, then, if the polyhedron of the system $Ax \leq b + \Delta b$ is not empty then it is bounded. In this case, the problem

$$\max\{(c, x) \mid Ax \leq b + \Delta b\}$$

will be solvable, evidently, for each c. Then the dual problem

$$\min\{(b, u) \mid A^T u = c, \qquad u \geq 0\}$$

will be solvable also; therefore,

$$M^* = \{u \geq 0 \mid A^T u = c\} \neq 0.$$

These considerations justify the sufficiency of the condition of being multifold for the system $\{a_j\}_1^m$ for the problem L to be improper of the first order (with incompatible system of restrictions).

This condition may be weakened to the condition of being multifold for the system of vectors $\{a_j, -c\}_1^m$. In this case, we provide boundedness of the polyhedron of the system of inequalities

$$Ax \leq b + \Delta b, \qquad (c, x) \geq \alpha \tag{22.3}$$

for apposite Δb and α. This, evidently, yields solvability of the problem $\max_{Ax \leq b + \Delta b}(c, x)$. It is clear that if Δb is such that the system $Ax \leq b + \Delta b$ is compatible, then the choice of α which makes system (22.3) compatible is guaranteed.

Remark 22.1. If the initial problem is written in form (22.1), then in the above consideration we must consider the systems of vectors $\{a_j, -e_i\}_{1,1}^{m,n}$ and $\{a_j, -e_i, -c\}_{1,1}^{m,n}$ respectively to be multifold. Here $\{e_i\}_1^n$ are the unit vectors of the space x.

23.   INFORMATIVE INTERPRETATION OF IMPROPER PROBLEMS OF LINEAR PROGRAMMING

The practice of modelling the concrete problems of production planning and optimal constructing had collided with the situation of contradictive model. For example, in the *LP* model we may obtain the inconsistent system of restrictions (the system of linear inequalities). If we operate with such models, we obtain (as a result) the inconsistency (for example: $1 \leq 0$). We shall clarify this situation taking as an example the incompatible system of linear equations. The equations of such system can be always combined so that, as a result, we obtain the inconsistent relation $1 = 0$. What is the reason of arising the inconsistent models in optimal planning? We shall do, at first, some remarks. The principle of optimal solution in mathematical methods in economics is to choose from the set of admissible vectors the only vector which is optimal in the sense of a certain criterion (the maximal income, the minimum of labour expenditure and so on). If we consider the vector $x^T = [x_1, \ldots, x_n] \in \mathbb{E}_n$; which describes the plan, then the admissible domain can be represented as intersection of the two sets

$\mathbb{E}_n \supset M$ is the domain of technological and resource possibilities of the object of planning (control, projecting);

N is the domain of requirements, characteristics, estimates to which the model object must satisfy.

The admissible plan (project) x is the object which belongs both the set M and N; i.e., $x \in M \cap N$.

However, the most typical situation is when the requirements for the object we plan contradict with the technological and resource possibilities. Mathematically this situation can be expressed as $M \cap N = \varnothing$. The solution of this problem (in the informative sense) may be seen in minimal widening sets M and N which provides the non-emptiness of $M \cap N$. From the one hand this denotes widening the technological and resource possibilities; from the other hand — the weakening requirements to the object we plan.

Investigating the problems of theoretical economic problems is connected with necessity of scientific justification of such procedures of correction.

The sources of contradictions in models may be following: resource deficit, the lack of reserves of production capacity, inaccuracy of economic information, the account of contradictory conditions, the account of negative action of production on the surroundings and so on.

The contradictory models are rather tipical for practice. In simple models the overcome of difficulties is not so hard (weakening the restrictions, correction of the information and so on). Investigation of more complicated problems which represent more complicated situation leads to principal necessity of account of improper models. This means that we have to develop

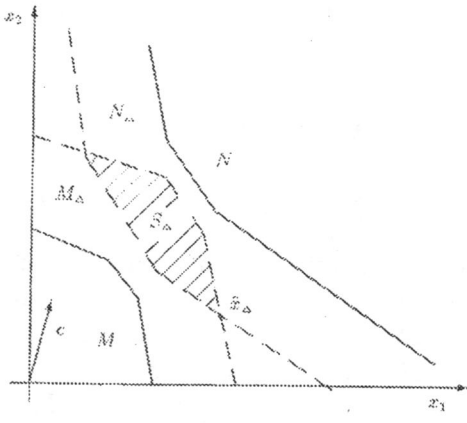

Figure 5

the theory of such problems (in particular, of duality), the methods of numerical analysis and its software.

We shall give below the informative interpretation of improperness of 1-3 orders of LP problems. Consider, first, model (13.8) and its demonstration in Figure 1 for the case $S = M \cap N = \varnothing$.

In the previous economic interpretation of model (13.8) the polyhedrons of possibilities and requirements. M and N were situated in contradictory state $M \cap N = \varnothing$. Diminishing the requirements and widening possibilities, i.e., introducing the correcting variations Δb and Δd lead us to not empty set S_Δ of admissible vectors x, where we can choose the optimal vector $\tilde{x}_\Delta$. This is illustrated in Figure 5. We see here the Δ-widering M_Δ and N_Δ of the sets M and N with non-empty intersection $S_\Delta := M_\Delta \cap N_\Delta$, where

$$M_\Delta := \{x \geq 0 \mid Ax \leq b + \Delta b\}, \qquad \Delta b \geq 0,$$
$$N_\Delta := \{x \geq 0 \mid Bx \geq d - \Delta d\}, \qquad \Delta d \geq 0.$$

In model (13.8), for $S = \varnothing$, the sufficient criterion for improperness of the 1st order will hold if the system of vectors $\{a_j, -b_s\}$ is multifold. Here $\{a_j\}$ are the rows of the matrix A; and $\{b_s\}$ are the rows of the matrix B. This situation is pictured in Figure 5.

If the LP problem, for example, given in form (13.5) has the improperness of the second order, there arises the stability problem. This is the problem on continuity of the optimum function $\tilde{f}(c)$ with respect to c (as the fragment of informative component of model (13.5)). If the problem L is solvable and its optimal set $\tilde{M}$ is bounded, then the function $\tilde{f}(y)$ will

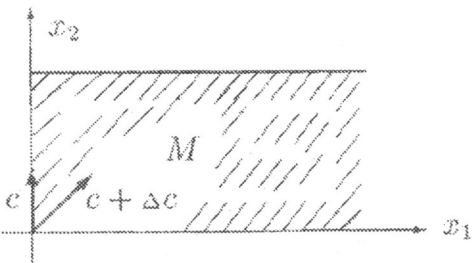

Figure 6

be continuous in the point $y = c$ (more detailed in Chapter 5). If $\tilde{M}$ is unbounded, then there exist extremely small variations Δc of the vector c so that the problem

$$L_{\Delta c} : \quad \max\{(c + \Delta c, x) \mid Ax \leq b, \; x \geq 0\} \tag{23.1}$$

will be unsolvable, i.e. opt $(23.1) = +\infty$. This denotes that L is the improper problem of the second order. Figure 6 illustrates this situation.

In this situation the function $\tilde{f}(y)$ has discontinuity in the point $y = c$.

The characterization given above of $IPLP$ of the second order is purely mathematical. As for its economical interpretation, the problem is to justify the actual economy-mathematical models (in its exact setting) with the unbounded optimal set. If such models were existed, then using the factor of approximate setting of the information which define the model we should obtain the $IPLP$ of the second order.

As for $IPLP$ of the third order, they can be characterized similar to $IPLP$ of the second order in terms of stability via violation of continuity of the optimum function

$$\tilde{f}(y, z) := \max\{(y, x) \mid Ax \leq z, \; x \geq 0\} \tag{23.2}$$

in the point $[c, b]$. Here c and b are taken from the solvable problem

$$L : \max\{(c, x) \mid Ax \leq b, \qquad x \geq 0\}.$$

More exact sense is as follows: if the optimal sets of L and L^* are unbounded, then there exist arbitrarily small variations Δc and Δd such that the problem

$$\max\{(c - \Delta c, x) \mid Ax \leq b + \Delta b, \; x \geq 0\} \tag{23.3}$$

will be $IPLP$ of the third order. So, an improper problem of the third order may exist because of approximate setting the vectors c and b.

Note that the characterization of $IPLP$ of the first, second, and third orders was realized from the positions of those economic situations which generate the improper LP problems.

24. METHODS OF CORRECTION OF IMPROPER PROBLEMS OF LINEAR PROGRAMMING: GENERAL APPROACHES

An improper LP problem can be approximated by a proper problem (i.e., solvable). The ways of such approximation may be various following the various optimality criterions. One of the most general schemes of optimal correction (approximation) of $IPLP$ is the scheme of *parameter imbedding*. In this scheme the optimization problem C is imbedding in a parameter class of problems $\{C_\Delta\}$. Here Δ is the parameter vector from a certain set Ω_0. The imbedding C in $\{C_\Delta\}$ denotes that for a certain $\Delta_0 \in \Omega_0$ the problems C and C_{Δ_0} coincides. We define the set

$$\Omega = \{\Delta_0 \in \Omega_0 \mid C_\Delta \text{ has a certain property } \omega\}$$

Taking the value of parameter Δ from Ω we obtain the problem C_Δ with the property ω that we are interested. This property may be solvability (which in the linear case coincides with properness). The problem C_Δ may be considered as correction of the problem C. However, if we introduce the function of *quality of correction* $\varphi(\Delta)$, then we can set the problem of *optimal correctness* in the form

$$\min\{\varphi(\Delta) \mid \Delta \in \Omega\}. \tag{24.1}$$

This problem can be setted in more complicated form, for example,

$$\min_{\Delta_1 \in \Omega_1} \max_{\Delta_2 \in \Omega_2} \varphi(\Omega), \tag{24.2}$$

where $\Delta = [\Delta_1, \Delta_2] \in \Omega = \Omega_1 \times \Omega_2$. This setting corresponds to the play situation when correction is realized by different "person" with different possibilities (relatively the fragments of the vector Δ) and opposite interests.

24.1. Correction of an improper problem of linear programming by its informative component

We shall illustrate now the method of optimal correction respectively scheme (24.1) considering some examples of the models in the situation when they are improper. Firstly, we take model (22.1):

$$L: \quad \max\{(c, x) \mid Ax \leq b, \ x \geq 0\}, \tag{24.3}$$

without clarifying the order of its improperness. The following class of problems with increments $\Delta = [\Delta c, \Delta b]$

$$\max\{(c - \Delta c, x) \mid Ax \leq b + \Delta b, \ x \geq 0\} \tag{24.4}$$

we shall consider together with this model. We set $\Omega_0 = \{\Delta \geq 0\}$. The set $\Omega \subset \Omega_0$ of increments $\Delta \geq 0$ which provide solvability of problem (24.4) is the set of those $\Delta c \geq 0$ and $\Delta b \geq 0$ which provides compatibility of systems

$$Ax \leq b + \Delta b, \qquad x \geq 0 \tag{24.5}$$

and

$$A^T u \geq c - \Delta c, \qquad u \geq 0 \tag{24.6}$$

with respect to x and u respectively. This means that

$$M_{\Delta b} := \{x \geq 0 \mid Ax \leq b + \Delta b\} \neq \varnothing,$$
$$M^*_{\Delta c} := \{u \geq 0 \mid A^T u \geq c - \Delta c\} \neq \varnothing.$$

If we take the function $\varphi(\Delta)$ in the form $\varphi(\Delta) = \|\Delta c\|^2 + \|\Delta b\|^2$, then problem (24.1) will be written in the form

$$\min\{\|\Delta b\|^2 + \|\Delta c\|^2 \mid Ax \leq b + \Delta b, \quad x \geq 0;$$
$$A^T u \geq c - \Delta c, \quad u \geq 0; \quad [\Delta b, \Delta c] \geq 0\}. \tag{24.7}$$

This problem falls into two independent problems

$$\min\{\|\Delta b\|^2 \mid Ax \leq b + \Delta b, \ x \geq 0, \ \Delta b \geq 0\}, \tag{24.8}$$

$$\min\{\|\Delta c\|^2 \mid A^T u \geq c - \Delta c, \ u \geq 0, \ \Delta c \geq 0\}. \tag{24.9}$$

Both these problems are solvable problems of square programming. If $\overline{\Delta b} = \arg(24.8)$ and $\overline{\Delta c} = \arg(24.9)$, then

$$L_{\overline{\Delta}}: \quad \max\{(c - \overline{\Delta c}, x) \mid Ax \leq b + \overline{\Delta b}, \ x \geq 0\} \tag{24.10}$$

is optimal problem corrected for (24.3). Problems (24.8) and (24.9) are the problems of square approximation of system of linear inequalities $Ax \leq b$, $x \geq 0$ and $A^T u \geq c$, $u \geq 0$. These problems can be written briefly in the form

$$\min_{x \geq 0} \|(Ax - b)^+\|^2, \qquad \min_{u \geq 0} \|(c - A^T u)^+\|^2. \qquad (24.11)$$

Here $+$ denotes the positive cut-off of the vector.

The following evident statement holds $\overline{\Delta b} = 0$, $\overline{\Delta c} = 0$ corresponds for solvability of L.

$\overline{\Delta b} \neq 0$, $\overline{\Delta c} = 0$ corresponds for improperness of L of the first order;

$\overline{\Delta b} = 0$, $\overline{\Delta c} \neq 0$ corresponds to improperness of L of the second order.

$\overline{\Delta b} \neq 0$, $\overline{\Delta c} \neq 0$ corresponds to improperness of L of the third order.

We consider now model (13.8), i.e.,

$$\min\{(c, x) \mid Ax \leq b, \ Bx \geq d, \ x \geq 0\}, \qquad (24.12)$$

whose geometric illustration is given in Figure 1. Economic interpretation of this model defines the order of its improperness in the case $M \cap N = \varnothing$; namely, it is improperness of the first order (Figure 2). Hence it follows that approximation (correction) of problem (24.12) is reduced to approximation of its system of restrictions

$$Ax \leq b, \qquad Bx \geq d, \qquad x \geq 0.$$

Analogously to (24.8) it will be as follows:

$$\min\{\|\Delta b\|^2 + \|\Delta d\|^2 \mid Ax \leq b + \Delta b, \ Bx \geq d - \Delta d, \ x \geq 0, \ [\Delta b, \Delta d] \geq 0\}.$$

We have taken above the square function $\varphi(\Delta) = \|\Delta\|^2$ as the function of correction of quality. This function may be taken linear if we may justify this. We take the function $\varphi(\Delta) = (\bar{u}, \Delta b) + (\bar{v}, \Delta d)$ as $\varphi(\Delta)$ with the following sense of the vectors $\bar{u} \geq 0$ and $\bar{v} \geq 0$. The vector $\bar{u}$ is the price vector for resources; the additional part is Δb; $\bar{v}$ is the price vector for the product. Then $(\bar{u}, \Delta b)$ are the losses connected with buying additional resources; $(\bar{v}, \Delta d)$ are the losses connected with unsaled production Δd. The general losses is the sum $(\bar{u}, \Delta b) + (\bar{v}, \Delta d)$ which must be minimized:

$$\min\{(\bar{u}, \Delta b) + (\bar{v}, \Delta d) \mid Ax \leq b + \Delta b, \ Bx \geq d - \Delta d, \ x \geq 0, \ [\Delta b, \Delta d] \geq 0\}.$$

Thus, the problem of correction is written in the form of *LP* problem. As in the case (24.11), we can write this problem as follows:

$$\min_{x \geq 0} \{ (\bar{u}, (Ax - b)^+) + (\bar{v}, (d - Bx)^+) \}. \tag{24.13}$$

Corrected initial problem (24.12) can be written in the form

$$\min \{ (c, x) \mid x \in \text{Arg} \ (24.13) \}.$$

From informative point of view the correction of the vector b (for example) in problem (24.3) by means of increment Δb may be not free. It may be *dependent*, suppose in the form $\Delta b = t \overline{\Delta b}$, where $\overline{\Delta b}$ is a fixed vector of coordinate proportions; t is a free non-negative variable. Then the problem of correction (24.8) for replaced criterion $\varphi(\Delta) = (\bar{u}, \Delta b)$ becomes as follows:

$$\min \{ t(\bar{u}, \overline{\Delta b}) \mid Ax \leq b + t \overline{\Delta b}, \ x \geq 0, \ t \geq 0 \}. \tag{24.14}$$

From the sense: $\bar{u} > 0$ $\overline{\Delta b} = [\overline{\Delta b_1}, \dots, \overline{\Delta b_m}]^T > 0$; therefore, $(\bar{u}, \overline{\Delta b}) > 0$. The j-inequality of the system $Ax \leq b + t \overline{\Delta b}$ we divide by $\overline{\Delta b_j}$ $(j = 1, \dots, m)$ and exclude the variables x_i $(i = 1, \dots, n)$ of the vector x from the system by the Fourier method (see Section 11). As a result, from the system $t \geq \beta_s$ $(s = 1, \dots, l)$, which we obtain, we determine the optimal value t_{opt} in problem (24.14) following the relation $t_{\text{opt}} = \max_{(s)} \beta_s^+$.

Correction of *LP* problem, suppose of (24.3), with respect to vectors b and c is sufficient to provide solvability of this problem. This was the reason why we have considered this case in details. The second reason was that the correction with respect to these fragments of informative component is tied up with classification of *IPLP*. In whole, we can set the problem on optimal correction of the model L with respect to each part of complete system of information $\{ c_i, b_j, a_{ij} \}$ on the problem L. Here $[c_1, \dots, c_n]^T = c$, $[b_1, \dots, b_m]^T = b$, $[a_{ij}]_{1,1}^{m,n} = A$. This problem is rather difficult and we do not consider it here.

24.2. Other methods of optimal correction

As we may see from the previous section the most important type of optimal correction of problem L (suppose of (24.3)) is the approximation of incompatible systems of linear inequalities (we mean the systems of restrictions for problems L and L^*). This approximation can be realized in various ways. One of them is the method of least squares: $\min_{x \geq 0} \| (Ax - b)^+ \|^2$.

We consider now the ways of approximation (optimal correction) of the system

$$Ax \leq b, \qquad x \geq 0, \tag{24.15}$$

setting in the basis the construction of Pareto and successive optimization (see Section 21).

We suppose that system (24.15) is divided orbitrarily onto subsystems

$$A_j x \leq b^j, \qquad j = 0, 1, \ldots, m_0.$$

In this case, the condition $x \geq 0$ we shall adjoin to the system $A_0 x \leq b^0$ and set

$$M_0 = \{x \geq 0 \mid A_0 x \leq b_0\} \neq 0.$$

The restrictions of the subsystems $A_0 x \leq b_0$ we consider as *directive* restrictions; all other restrictions are *auxiliary* restrictions, ordered (possibly) by their importance. Choosing an arbitrary system of norms $\| \cdot \|_j$ in the spaces of vectors b^j we form the functions

$$f_j(x) = \|(A_j x - b_j)^+\|_j, \qquad j = 1, \ldots, m_0.$$

Let $\mathcal{F} = \{f_j(x)\}_1^{m_0}$ and p be an arbitrary order of the functions $f_j(x)$. We can write the two problems

$$\min_{\pi} \{\mathcal{F} \mid A_0 x \leq b^0, \ x \geq 0\}, \tag{24.16}$$

$$\min_{p} \{\mathcal{F} \mid A_0 x \leq b^0, \ x \geq 0\}. \tag{24.17}$$

The first problem is the problem of Pareto optimization with respect to the system $\mathcal{F}$; the second problem is the problem of subsequence optimization with respect to the system $\mathcal{F}$ also for a certain order $p = (j_1, \ldots, j_{m_0})$. For example, $p = (m_0, \ldots, 1)$.

If problem (24.3) is *IPLP* of the first order, then we can write the following aproximation problems for these two problems:

$$\max\{(c, x) \mid x \in \text{Arg}\,(24.16)\}, \tag{24.18}$$

$$\max\{(c, x) \mid x \in \text{Arg}\,(24.17)\}. \tag{24.19}$$

These two problems have similar reduction (by the form)

$$\max \left\{ (c, x) - \sum_{j=1}^{m_0} R_j \|(A_j x - b^j)^+\|_j \ \middle| \ A_0 x \leq b^0, \ x \geq 0 \right\}. \tag{24.20}$$

The basis of this reduction is the method of exact penalty functions (see Section 20) and Theorem 21.5 from Section 21. More detailed setting of these problems will be given in Chapter 4, where the duality problems for Pareto and subsequent optimization problems will be considered.

In conclusion we note that for inconsistent optimization models there exist a lot of methods of optimal corrections. Some of them were considered above. However, concrete realization of the methods depends on the problem we model, on the practical sense of the problem, on the order of positions which characterize the object: the hierarchy of resources, equipment and the other ingradients of production. Moreover, the method we must choose. In whole, different approaches to forming the correction methods can be united in the ideology of *compromise models* for multicriterion inconsistent optimization problems.

25. DUALITY: THE MAIN THEOREM

We shall consider here the apparatus of duality for improper LP problems. In particular, various formulations of the duality theorem and their approximation sense are examined.

Here we have the two mutually dual LP problems

$$L : \quad \max\{(c, x) \mid Ax \le b, \ x \ge 0\}; \tag{25.1}$$

$$L^* : \quad \min\{(b, u) \mid A^T u \ge c, \ u \ge 0\}. \tag{25.1*}$$

The rule $(*)$ in forming the problem L^* by the problem L has the property of duality. This means that if we apply this rule to L^* we obtain $L : (L^*)^* = L$. The problems L and L^* are connected by the following theorem.

Duality Theorem (Theorem 15.2). If one of the problems (25.1), (25.1*) is solvable, then the other problem is solvable too. Moreover, their optimal values coincide: $\operatorname{opt} L = \operatorname{opt} L^*$.

Such pairs of mutually dual problems we shall call *regularly dual*.

25.1. The scheme of forming a pair of mutually dual problems C and $C^{\#}$ corresponding to the case of an improper problem L

Here we shall use a certain scheme τ and form the two proper problems C and $C^{\#}$ by the initial pair of dual problems L and L^*. The problems C

and $C^\#$ will be connected by the regular duality and they will correspond to L and L^* respectively. The scheme is as follows:

$$
\begin{array}{ccc}
L & \xrightarrow{\ \tau\ } & C \\[2pt]
(*) \Big\downarrow\Big\uparrow & & \Big\downarrow\Big\uparrow (\#) \\[2pt]
L^* & \xrightarrow{\ \tau\ } & C^\#
\end{array}
$$

Scheme 2

Note that there are may schemes which satisfy the above properties. In order to choose the concrete scheme τ we should impose certain natural requirements, for example: the scheme must be sufficiently general and have possibility of various approximations of improper LP problems. The scheme must also enrich the duality. These requirements were taken into account when constructing the scheme τ considered below.

We fix an arbitrary cut of the matrix A onto the submatrices A_j, $j = 0, 1, \ldots, m_0$ and B_i, $i = 0, 1, \ldots, n_0$ in horizontal and vertical directions respectively.

$$
A = \begin{bmatrix} A_0 \\ A_1 \\ \vdots \\ A_{m_0} \end{bmatrix} = [B_0, B_1, \ldots, B_{n_0}].
$$

The correspondent cuts of vectors b, u, c and x are as follows:

$$
b^T = [b^0, b^1, \ldots, b^{m_0}], \qquad u^T = [u^0, u^1, \ldots, u^{m_0}];
$$
$$
c^T = [c^0, c^1, \ldots, c^{n_0}], \qquad x^T = [x^0, x^1, \ldots, x^{n_0}].
$$

We shall give to chosen submatrices the meaning of empty values $\varnothing$ if it is necessary. For example, if $A_1 = \varnothing$, $i = 1, \ldots, n_0$, then $A = A_0$ and so on.

We suppose that $\|\cdot\|_j$ and $\|\cdot\|_i$ are arbitrary monotone norms in the spaces of vectors u^j and x^i, $j = 0, 1, \ldots, m_0$, $i = 0, 1, \ldots, n_0$; $\|\cdot\|_j^*$ and $\|\cdot\|_i^*$ are the conjugate norms which are assumed to be monotone also.

The norm $\|\cdot\|$ in the n-dimensional space $\mathbb{E}_n$ is *monotone* if

$$
0 \le x \le y \quad \Longrightarrow \quad \|x\| \le \|y\|,
$$

where $x, y \in \mathbb{E}_n$. The following norms are monotone:

$$\|x\|_0 = \max_{1 \le i \le n} |x_i|, \quad \|x\|_1 = \sum_{i=1}^{n} |x_i|, \quad \|x\|_2 = (\sum_{i=1}^{n} x_i^2)^{1/2}.$$

The norms $\|x\|_0$ and $\|x\|_1$ are piecewise linear. An arbitrary piecewise linear norm $\|x\|_l$ is given by the expression $\max_{(k)} |(c_k, x)|$. In this case, among the finite collection of vectors $\{c_k\}$ there are n linearly independent vectors. The property of being monotone for the norm $\|\cdot\|_l$ denotes that $c_k \ge 0$, $\forall k$ (the algebra of piecewise linear functions is considered in Chapter 7).

Suppose $R_j > 0$, $j = 1, \ldots, m_0$, $r_i > 0$, $i = 1, \ldots, n_0$ is the system of positive parameters. We formulate the problems

$$C: \quad \sup \Big\{ (c, x) - \sum_{j=1}^{m_0} R_j \|(A_j x - b^j)^+\|_j \ \Big| \ A_0 x \le b^0, \quad x \ge 0,$$

$$\|x^i\|_i \le r_i, \quad i = 1, \ldots, n_0 \Big\}; \tag{25.2}$$

$$C^{\#}: \quad \inf \Big\{ (b, u) + \sum_{i=1}^{n_0} r_i \|(c^i - B_i^T u)^+\|_i^* \ \Big| \ B_0^T u \ge c^0, \quad u \ge 0,$$

$$\|u^j\|_j^* \le R_j, \quad j = 1, \ldots, m_0 \Big\}; \tag{25.2\#}$$

Remark 25.1. Positiveness of R_j and r_i may be weakened to non-negativeness. If one of the numbers R_j, for example R_1, attains the zero value, this denotes that restrictions correspondent to the submatrix A_1 are removed from the initial problem L, If $r_1 = 0$, then we remove the submatrix B_1 with the vector C' from the problem L. Such operations of shortening L must be forseen (especially when we realize the programm package which provides the numerical analysis of improper problems of linear programming).

Remark 25.2. Problems (25.2) and (25.2$^{\#}$) for $A_j = \varnothing$, $j = 1, \ldots, m_0$, $B_i = \varnothing$, $i = 1, \ldots, n_0$ become problems (25.1) and (25.1*).

Subsystems $A_0 x \le b^0$, $x \ge 0$ and $B_0^T u \ge c^0$, $u \ge 0$ of restriction systems in problems (25.1) and (25.1*) correspondent to submatrices A_0 and B_0 (which may attain the values $\varnothing$) are supposed to be compatible. This assumption is not overloaded since we may always select the compatible systems from incompatible systems. This corresponds to selection of directive restrictions.

Remark 25.3. From the form of problems C and $C^{\#}$ we see that the scheme of their construction from the problems L and L^* is unique (the scheme τ).

Remark 25.4. We suppose that $(\#)$ is the rule: $C\,C \xrightarrow{(\#)} C^{\#}$. It is easy to see that (taking into account the property $(\|\cdot\|^*)^* = \|\cdot\|$, where $\|\cdot\|$ is any norm of a finite dimensional space) this rule is mutual, i.e., $(C^{\#})^{\#} = C$.

Remark 25.5. In problems (25.2) and (25.2$^{\#}$) the goal functions, which we denote as $f_R(x)$ and $f_r^{\#}(u)$ are concave and convex respectively. Therefore, the problems C and $C^{\#}$ are the problems of convex programming, taking into account convexity of inequalities

$$\|x^i\|_i \leq r_i, \quad i = 1,\ldots,n_0, \qquad \|u^j\|_j^* \leq R_j, \quad j = 1,\ldots,m_0.$$

If in ptoblem (25.2) all $\|\cdot\|_j$ and $\|\cdot\|_i$ are norms of type $\|\cdot\|_0$ and $\|\cdot\|_1$, we shall call it the *l-problems*. As problem (25.2) is *l*-problem then problem (25.2$^{\#}$) is *l*-problem also and conversely.

Remark 25.6. If (25.2) is *l*-problem than C and $C^{\#}$ are convex piecewise linear and can be rewritten equivalently in the form of LP problems.

Remark 25.7. In (25.2) and (25.2$^{\#}$) we use operations sup and inf since max and min may be not attained. The examples of these situations are trivial.

25.2. Main theorem

The following statement holds.

Theorem 25.1. *For each*

$$\bar{x} \in M(r) := \{x \geq 0 \mid A_0 x \leq b^0,\ \|x^i\|_i \leq r_i, i = 1,\ldots,n_0\},$$

$$\bar{u} \in M^{\#}(R) := \{u \geq 0 \mid B_0^T \geq c^0,\ \|u^j\|_j^* \leq R_j,\ j = 1,\ldots,m_0\}$$

the inequality

$$f_R(\bar{x}) \leq f_r^{\#}(\bar{u})$$

holds.

Proof. Below we shall use the inequalities:

$$(u^j, (A_j x - b^j)^+) \leq \|u^j\|_j^* \, \|(A_j x - b^j)^+\|_j, \qquad j = 1, \ldots, m_0, \qquad (25.3)$$

$$(x^i, (c^i - B_i^T u)^+) \leq \|x^i\|_i \, \|(c^i - B_i^T u)^+\|_i^*, \qquad i = 1, \ldots, n_0, \qquad (25.4)$$

which follow for each x and u from the definition of conjugate norm. Taking into account (25.3) and conditions $\|u^j\|_j^* \leq R_j$, $j = 1, \ldots, m_0$, we have

$$f_R \leq (c, \bar{x}) - \sum_{j=1}^{m_0} \|\bar{u}^j\|_j^* \cdot \|(A_j \bar{x} - b^j)^+\|_j$$

$$\leq (c, \bar{x}) - \sum_{j=1}^{m_0} (\bar{u}^j, (A_j \bar{x} - b^j)^+).$$

As $A_j \bar{x} - b^j \leq (A_j \bar{x} - b^j)^+$, $j = 1, \ldots, m_0$, then

$$f_R(\bar{x}) \leq (c, \bar{x}) - \sum_{j=1}^{m_0} (\bar{u}^j, A_j \bar{x} - b^j).$$

This relation, the equality

$$\sum_{j=1}^{m_0} (\bar{u}^j, A_j \bar{x} - b^j) = (\bar{u}, A\bar{x} - b) - (\bar{u}^0, A_0 \bar{x} - b^0)$$

and inequalities $A_0 \bar{x} - b^0 \leq 0$, $\bar{u}^0 \geq 0$ yield

$$f_R(\bar{x}) \leq (c, \bar{x}) - (\bar{u}, A\bar{x} - b) + (\bar{u}^0, A_0 \bar{x} - b^0)$$

$$\leq (c, \bar{x}) - (\bar{u}, A\bar{x} - b) = (c - A^T \bar{u}, \bar{x}) + (b, \bar{u})$$

$$= (b, \bar{u}) + \sum_{i=1}^{n_0} (\bar{x}^i, c^i - B_i^T \bar{u}) + (\bar{x}^0, c^0 - B_0^T \bar{u}).$$

This, taking into account (25.4) and $c^0 - B_0^T \bar{u} \leq 0$, yields

$$f_R(\bar{x}) \leq (b, \bar{u}) + \sum_{i=1}^{n_0} \|\bar{x}^i\|_i \, \|(c^i - B_i^T \bar{u})^+\|_i^*.$$

As $\|\bar{x}^i\|_i \le r_i$, $i = 1, \ldots, n_0$, then, finally, we have

$$f_R(\bar{x}) \le (b, \bar{u}) + \sum_{i=1}^{n_0} r_i \|(c^i - B_i^T \bar{u})^+\|_i^* = f_r^\#(\bar{u}).$$

$\square$

Corollary 25.1. *If the vectors $\bar{x} \in M(r)$ and $\bar{u} \in M^\#(R)$ provide the equality $f_R(\bar{x}) = f_r^\#(\bar{u})$, then they are optimal for problems (25.2) and (25.2$^\#$) respectively, i.e., $\bar{x} \in \text{Arg}\,(25.2)$ and $\bar{u} \in \text{Arg}\,(25.2^\#)$.*

Now, we formulate the basic statement on regular duality which connects problems (25.2) and (25.2$^\#$).

Theorem 25.2. *Suppose that problem (25.2) is solvable and r_i, $i = 1, \ldots, n_0$ are such that*

$$\|x^i\|_i < r_i, \quad i = 1, \ldots, n_0, \quad A_0 x \le b^0, \quad x \ge 0 \qquad (25.5)$$

is compatible. Then problem (25.2$^\#$) is solvable and optimal values for problems (25.2) and (25.2$^\#$) coincide:

$$\text{opt}\,(25.2) = \text{opt}\,(25.2^\#).$$

Remark 25.8. As the problems (25.2) and (25.2$^\#$) are mutual, the following inverse theorem holds.

Solvability of (25.2$^\#$) and compatibility of the system

$$\|u^j\|_j^* < R_j, \quad j = 1, \ldots, m_0, \quad B_0^T u \ge c^0, \quad u \ge 0 \qquad (25.6)$$

yield solvability of (25.2) and coincidence of optimal values of (25.2$^\#$) and (25.2).

Remark 25.9. Compatibility of system (25.5) in the formulation of the duality theorem (or of system (25.6) in the formulation of the inverse theorem) is equivalent to the R_0-regularity (see Application A, Section A3). This condition provides for solvable problem of convex programming with restrictions of problem (25.2) the validity of the Kun–Takker theorem (see

Application A, Theorem 8). For certain particular norms $\|\cdot\|_j$, $\|\cdot\|_i$, for example, when these norms are piecewise linear, we can omit the rigorous inequalities $\|x^i\|_i < r_i$, $i = 1, \ldots, n_0$ in system (25.5). Further, we shall use this fact.

Before we prove the theorem, we describe an auxiliary construction and formulate some lemmas.

Problem (25.2) we rewrite in the equivalent form

$$\sup\Big\{(c, x) - \sum_{j=1}^{m_0} R_j \|t^j\|_j \;\Big|\; A_j x \le b^j + t^j, \quad j = 1, \ldots, m_0,$$

$$\|x^i\|_i \le r_i, \quad i = 1, \ldots, n_0, \quad A_0 x \le b^0, \quad [x, t] \ge 0\Big\}; \qquad (25.7)$$

here $t = [t^1, \ldots, t^{m_0}]^T$. The Lagrange function correspondent to this problem we take as follows

$$F(x, t; u, v) = (c, x) - \sum_{j=1}^{m_0} R_j \|t^j\|_j - \sum_{j=1}^{m_0} (u^j, A_j x - b^j - t^j)$$

$$- \sum_{i=1}^{n_0} v_i (\|x^i\|_i - r_i) - (u^0, A_0 x - b^0);$$

here $u^T = [u^0, u^1, \ldots, u^{m_0}] \ge 0$, $v^T = [v_1, \ldots, v_{n_0}] \ge 0$ are vectors of nonnegative Lagrange multipliers.

Lemma 25.1. *The following inequality holds*

$$F(x, t; u, v) = (b, u) + (c^0 - B_0^T u, x^0) + \sum_{i=1}^{n_0} v_i r_i + \sum_{j=1}^{m_0} \left[-R_j \|t^j\|_j + (t^j, u^j)\right]$$

$$+ \sum_{i=1}^{n_0} \left[(c^i - B_i^T u, x^i) - v_i \|x^i\|_i\right]. \qquad (25.8)$$

Proof. The proof of this lemma is purely computational. So, we have

$$F(x, t; u, v) = \sum_{i=1}^{n_0} (c^i, x^i) + \sum_{j=1}^{m_0} (u^j, b^j) - \sum_{j=1}^{m_0} R_j \|t^j\|_j + \sum_{j=1}^{m_0} (u^j, t^j)$$

$$-\sum_{j=1}^{m_0}(A_j^T u^j, x) - \sum_{i=1}^{n_0}v_i(\|x^i\|_i - r_i) - (u^0, A_0 x - b^0)$$

$$= (b, u) + \sum_{i=1}^{n_0}(c^i - B_i^T u, x^i) + (c^0 - B_0^T u, x^0) + \sum_{j=1}^{m_0}(u^j, t^j)$$

$$-\sum_{j=1}^{m_0}R_j\|t^j\|_j - \sum_{i=1}^{n_0}v_i(\|x^i\|_i - r_i)$$

$$= (b, u) + (c^0 - B_0^T u, x^0) + \sum_{i=1}^{n_0}v_i r_i + \sum_{j=1}^{m_0}[-R_j\|t^j\|_j + (t^j, u^j)]$$

$$+ \sum_{i=1}^{n_0}[(c^i - B_i^T u, x^i) - v_i\|x^i\|_i]$$

which was to be proved. $\square$

We set now $F_0(x^0, u) = (c^0 - B_0^T u, x_0)$, $F_1(t^j, u^j) = (t^j, u^j) - R_j\|t^j\|_j$, and $F_2(x^i, u, v_i) = (c^i - B_i^T u, x^i) - v_i\|x^i\|_i$.

Lemma 25.2. *The following relation holds*

$$\sup_{t^j \geq 0} F_1(t^j, u^j) = \begin{cases} 0, & \text{if } \|u^j\|_j^* \leq R_j, \\ +\infty, & \text{otherwise} \end{cases}.$$

Proof. As

$$(t^j, u^j) \leq \|t^j\|_j \|u^j\|_j^*, \tag{25.9}$$

then, for $\|u^j\|_j^* \leq R_j$, we shall have

$$F_1(t^j, u^j) \leq \|t^j\|_j(-R_j + \|u^j\|_j^*) \leq 0, \qquad \forall t^j \geq 0.$$

Taking into account $F_1(0) = 0$, we obtain the first part of the relation. Suppose now that $\|u^j\|_j^* > R_j$. For a certain $\tilde{t}^j \geq 0$, $\tilde{t}^j \neq 0$, we shall have the equality in (25.9). Setting $t^j = \alpha\tilde{t}^j$, we obtain

$$F_1(t^j, u^j) = \alpha\|\tilde{t}^j\|_j(-R_j + \|u^j\|_j^*) \to +\infty \qquad \text{for} \quad \alpha \to +\infty.$$

Thus, the second case is proved also. $\square$

Lemma 25.3. *The following relation holds*

$$\sup_{x^i \geq 0} F_2(x^i, u, v_i) = \begin{cases} 0, & \text{if } \|(c^i - B_i^T u)^+\|_i^* \leq v_i, \\ +\infty, & \text{otherwise} \end{cases}.$$

Proof. As

$$((c^i - B_i^T u)^+, x^i) \leq \|x^i\|_i \, \|(c^i - B_i^T u)^+\|_i^*, \qquad (25.10)$$

then for $\|(c^i - B_i^T u)^+\|_i^* \leq v_i$ we shall have

$$F_2(x^i, u, v_i) \leq \|x^i\|_i \left(-v_i + \|(c^i - B_i^T u)^+\|_i^*\right) \leq 0, \qquad \forall x^i \geq 0.$$

Taking into account $F_2(0) = 0$, we obtain the first case. Now, we suppose that

$$\|(c^i - B_i^T u)^+\|_i^* \geq v_i. \qquad (25.11)$$

For a certain $\bar{x}^i \geq 0$, $\bar{x}^i \neq 0$, inequality (25.10) becomes the equality

$$((c^i - B_i^T u)^+, \bar{x}^i) = \|\bar{x}^i\|_i \, \|(c^i - B_i^T u)^+\|_i^*. \qquad (25.12)$$

We form the vector $\tilde{x}^i$ from the vector $\bar{x}^i$ using the following rule. If a certain coordinate of the vector $(c^i - B_i^T u)^+$ is equal to zero, we take zero instead of corresponded coordinate of $\bar{x}^i$. In this case, if we replace $\bar{x}^i$ by $\tilde{x}^i$, the left-hand side in (25.12) will not change and the right-hand side will diminish, i.e.,

$$((c^i - B_i^T u)^+, \tilde{x}^i) \geq \|\tilde{x}^i\|_i \, \|(c^i - B_i^T u)^+\|_i^*. \qquad (25.13)$$

In this case, $\tilde{x}^i \neq 0$, otherwise all the coordinates of the vector $c^i - B_i^T u$ would be non-positive; and inequality (25.11) would be contradictorily: $0 > v_i \geq 0$.

We set $x^i = \alpha \tilde{x}^i$. Taking account of (25.13) and (25.11), we obtain

$$F_2(x^i, u, v_i) = \alpha \big[(c^i - B_i^T u, \tilde{x}^i) - v_i \|\tilde{x}^i\|_i\big] = \alpha \big[((c^i - B_i^T u)^+, \tilde{x}^i) - v_i \|\tilde{x}^i\|_i\big]$$

$$\geq \alpha \|\tilde{x}^i\|_i \left(\|(c^i - B_i^T u)^+\|_i^* - v_i \right) \to +\infty \qquad \text{for} \quad \alpha \to +\infty.$$

Therefore, we have proved the second case also.   □

Consider now the inequality system

$$\begin{cases} B_0^T u \geq c^0, \quad u \geq 0, \quad \|u^j\|_j^* \leq R_j, & j = 1, \dots, m_0; \\ \|(c^i - B_i^T u)^+\|_i^* \leq v_i, & i = 1, \dots, n_0. \end{cases} \tag{25.14}$$

Lemma 25.4. *The following relation holds*

$$\sup_{[x,t]\geq 0} F(x, t; u, v) = \begin{cases} (b, u) + \sum_{i=1}^{n_0} r_i v_i, & \text{if system (25.14) is compatible,} \\ +\infty, & \text{otherwise.} \end{cases}$$

Proof. The proof of this lemma we obtain from Lemmas 25.1–25.3 and from the evident relation

$$\sup_{x^0 \geq 0} F_0(x^0, u) = \begin{cases} 0, & \text{if } c^0 - B_0^T u \geq 0, \\ +\infty, & \text{otherwise.} \end{cases}$$

Lemma 25.4 yields the equality

$$\min_{[u,v]\geq 0} \max_{[x,t]\geq 0} F(x, t; u, v) = \min \left\{ (b, u) + \sum_{i=1}^{n_0} r_i v_i \; : \; u, v \text{ satisfy (25.14)} \right\}, \tag{25.15}$$

which shows that the left-hand side of this relation is equal to opt (25.2$^\#$) of problem (25.2$^\#$) because problem (25.15) is the equivalent form of problem (25.2$^\#$).

Now, we can finish the proof of Theorem 25.2. Namely, the theorem conditions allow us to apply the Kun–Takker theorem to problem (25.2) (see Appendix, Section A3, Theorem 8). By this theorem, we have

$$\max_{[x,t]\geq 0} \min_{[u,v]\geq 0} F(x, t; u, v) = \min_{[u,v]\geq 0} \max_{[x,t]\geq 0} F(x, t; u, v).$$

In this case, the set of optimal solutions of the left problem which coincides with the set of optimal solutions of problem (25.2) will coincide also with the set of optimal solutions of the right problem. As the left-hand side of the last equality is equal to opt (25.2) and the right-hand side, as was shown, is equal to opt (25.2#), then the theorem is proved. □

25.3. Corollaries from the main theorem

We shall give some corollaries of Theorem 25.2.

Corollary 25.2. *Suppose the conditions provide the regular duality for problems C and $C^\#$ (i.e. the problems are solvable and their optimal values coincide). Then, in order to find optimal vectors, we must find at least one solution of the system of linear and convex inequalities*

$$\begin{cases} A_0 x \leq b^0, & x \geq 0, & \|x^i\|_i \leq r_i, & i = 1, \ldots, n_0; \\ B_0 u \geq c^0, & u \geq 0, & \|u^j\|_j \leq R_j, & j = 1, \ldots, m_0; \\ \qquad\qquad f_r^\#(u) - f_R(x) \leq 0. \end{cases} \qquad (25.16)$$

We denote by $\tilde{f}(b, c)$ an optimal value of the problem C. It depends on the parameters b (the resource vector) and the vector c (the price vector). If conditions of Theorem 25.2 hold, $\tilde{f}(b, c)$, being the optimum function for $C^\#$ also, is concave with respect to b and convex with respect to c. This follows from equivalence of problem C and (25.7) and from equivalence of problem $C^\#$ and

$$\inf \Big\{ (b, u) + \sum_{i=1}^{n_0} r_i \|v_i\|_i^* \ \Big| \ c^i - B_i^T u \leq v_i, \quad i = 1, \ldots, n_0, $$

$$B_0^T u \geq c^0, \quad \|u^j\|_j \leq R_j, \quad j = 1, \ldots, m_0, \quad [u, v] \geq 0 \Big\} \qquad (25.17)$$

with the use of Theorem 14 (Appendix, Section A4).

We suppose now that $\widetilde{M}(b, c)$ and $\widetilde{M}^\#(b, c)$ are the optimal sets of the problems C and $C^\#$, respectively.

Corollary 25.3. *If in condition of Theorem 25.2, th sets $\widetilde{M}(b, c)$ and $\widetilde{M}^\#(b, c)$ are bounded, then*

1) for $s \geq 0$, we have

$$\frac{\partial \tilde{f}(b, c)}{\partial s} = \lim_{t \to +\infty} \frac{\tilde{f}(b + ts, c) - \tilde{f}(b, c)}{t} = \min\{(\tilde{u}, s) : \tilde{u} \in \widetilde{M}^\#(b, c)\};$$

2) for $l \leq 0$, we have

$$\frac{\partial \tilde{f}(b,c)}{\partial l} = \lim_{t \to +\infty} \frac{\tilde{f}(b, c+tl) - \tilde{f}(b,c)}{t} = \max\{(\tilde{x}, -l) : \tilde{x} \in \widetilde{M}(b,c)\}.$$

In particular, if $\widetilde{M}(b,c) = \{\tilde{x}\}$ and $\widetilde{M}^{\#}(b,c) = \{\tilde{u}\}$, then

$$\frac{\partial \tilde{f}(b,c)}{\partial s} = (\tilde{u}, s), \qquad \frac{\partial \tilde{f}(b,c)}{\partial l} = -(\tilde{x}, l).$$

For $s = [0, \ldots, s_j = 1, 0, \ldots, 0]$, $l = [0, \ldots, l_i = -1, 0, \ldots, 0]$, we obtain

$$\frac{\partial \tilde{f}(b,c)}{\partial b_j} = \tilde{u}_j, \qquad j = 1, \ldots, m,$$

$$\frac{\partial \tilde{f}(b,c)}{\partial(-c_i)} = -\tilde{x}_i, \qquad i = 1, \ldots, n,$$

where $\tilde{x}^T = [\tilde{x}_1, \ldots, \tilde{x}_n]$, $\tilde{u}^T = [\tilde{u}_1, \ldots, \tilde{u}_m]$.

This corollary is proved if we use Theorem 14 from Appendix and representation of problems C and $C^{\#}$ in forms (25.7) and (25.17) respectively. Also we use the equality $\tilde{f}(b,c) = \tilde{f}^{\#}(b,c)$, where $\tilde{f}^{\#}(b,c)$ is the optimal function for the problem $C^{\#}$ (or problem (25.17), which is similar).

26. SPECIAL REALIZATIONS OF DUALITY

We shall write here certain particular cases of problems (25.2) and (25.2$^{\#}$) and formulate the duality theorem for these cases.

Suppose in (25.2) and (25.2$^{\#}$) A_j is the j–row and B_i is the i–column of the matrix A (therefore , $A_0 = \varnothing$, $B_0 = \varnothing$). The problems (25.2) and (25.2$^{\#}$) will be as follows:

$$C_1 : \max\{(c, x) - (R, (Ax - b)^{+} \mid 0 \leq x \leq r\}, \tag{26.1}$$

$$C_1^{\#} : \min\{(b, u) + (r, (c - A^T u)^{+}) \mid 0 \leq u \leq R\} \tag{26.1$^{\#}$}$$

where $R = [R_1, \ldots, R_m]^T$, $r = [r_1, \ldots, r_n]^T$. In (26.1) and (26.1$^{\#}$) we write max and min instead of inf and sup since they are accessible.

Theorem 26.1. *Problems* (26.1) *and* (26.1#) *are solvable for each* $R \geq 0$ *and* $r \geq 0$; *their optimal values coincide.*

Suppose now that $A_0 = \varnothing$, $B_0 = \varnothing$, $m_0 = 1$, $n_0 = 1$, so that $A_1 = B_1 = A$. In this case we obtain the following realization for problems (25.2) and (25.2#):

$$\max\left\{ (c, x) - R_0\|(Ax - b)^+\|_p \;\middle|\; \|x\|_q \leq r_0, \; x \geq 0 \right\}, \qquad (26.2)$$

$$\min\left\{ (b, u) + r_0\|(c - A^T u)^+\|_q^* \;\middle|\; \|u\|_p^* \leq R_0, \; u \geq 0 \right\}. \qquad (26.2^{\#})$$

Here $\|\cdot\|_p$ and $\|\cdot\|_q$ are arbitrary monotone norms in the spaces $\mathbb{E}_m$ and $\mathbb{E}_n$, respectively; $R_0 \geq 0$, $r_0 \geq 0$. For this pair of problems the following theorem holds.

Theorem 26.2. *Problems* (26.2) *and* (26.2#) *are solvable always and their optimal values coincide.*

If (25.1) is an improper problem of the first order, then we can set $B_i = \varnothing$, $i = 1, \ldots, n_0$. Then $B_0 = A$ and problems (25.2) and (25.2#) can be written as follows:

$$\sup\left\{ (c, x) - \sum_{j=1}^{m_0} R_j\|(A_j x - b^j)^+\|_j \;\middle|\; A_0 x \leq b^0, \; x \geq 0 \right\}, \qquad (26.3)$$

$$\inf\left\{ (b, u) \;\middle|\; A^T u \geq c, \; u \geq 0, \; \|u^j\|_j^* \leq R_j, \; j = 1, \ldots, m_0 \right\}. \qquad (26.3^{\#})$$

In application to these problems Theorem 25.2 will have the following form.

Theorem 26.3. *Solvability of* (26.3) *yields solvability of* (26.3#). *Moreover, their optimal values coincide. Conversely, if* (26.3#) *is solvable and the system of inequalities*

$$A^T u \geq c, \qquad u \geq 0, \qquad \|u^j\|_j^* < R_j, \qquad j = 1, \ldots, m_0$$

is compatible, then (26.3) *is solvable also and the optimal values of the problems coincide.*

Another variant of the duality theorem for these two problems can be written as follows.

Theorem 26.4. *Suppose the restriction system in (26.3#) is compatible (by Theorem 25.1, this provides boundedness of optimal value of problem (26.3)) and the upper bound in problem (26.3) is accessible (therefore, this problem is solvable). Then problem (26.3#) is solvable also and the optimal values of (26.3) and (26.3#) coincide.*

Note that if $A_0 = \varnothing$, then the system of restrictions in problem (26.3#) defines the bounded set if this system is solvable. Therefore, the lower bound in (26.3#) is accessible. In this case, the inverse variant of the duality theorem will be as follows.

Theorem 26.5. *If the system of inequalities*

$$A^T u \geq c, \qquad u \geq 0, \qquad \|u^j\|_j^* < R_j, \qquad j = 1, \ldots, m_0$$

is compatible, then both problem (26.3#) and the problem

$$\max\left\{(c, x) - \sum_{j=1}^{m_0} R_j\|(A_j x - b^j)^+\|_j \ \Big| \ x \geq 0\right\} \tag{26.4}$$

are solvable and their optimal values coincide.

If, in addition, $m_0 = 1$, i.e., $A_1 = A$ then problems (26.3) and (26.3#) will become as follows:

$$\max\{(c, x) - R_0\|(Ax - b)^+\|_p \ | \ x \geq 0\} \tag{26.5}$$

$$\min\{(b, u) \ | \ A^T u \geq c, \ u \geq 0, \ \|u\|_p^* \leq R_0\} \tag{26.5#}$$

where $R_0 > 0$, $\|\cdot\|_p$ is an arbitrary norm from the space $\mathbb{E}_m$. The below theorem will be the formulation of the above theorem for this case.

Theorem 26.6. *If the system of inequalities*

$$A^T u \geq c, \qquad u \geq 0, \qquad \|u\|_p^* < R_0$$

is compatible (for this purpose it is necessary and sufficient that $R_0 > \min\{\|u\|_p^* \ | \ A^T u \geq c, \ u \geq 0\}$), then each of problems (26.5) and (26.5#) is compatible. In this case, their optimal values coincide.

We consider now the case when (25.1) is an improper problem of the second order. Then we can set $A_j = \varnothing$, $j = 1, \ldots, m_0$; therefore, $A_0 = A$. The following two dual problems will correspond to this case

$$\sup\{(c, x) \mid Ax \le b, \ x \ge 0, \ \|x^i\|_i \le r_i, \ i = 1, \ldots, n_0\}, \qquad (26.6)$$

$$\inf\left\{(b, u) + \sum_{i=1}^{n_0} r_i \|(c^i - B_i^T u)^+\|_i^* \ \bigg| \ B_0^T u \ge c_0, \ u \ge 0\right\}. \qquad (26.6^\#)$$

The following two analogs of Theorems 26.3 and 26.4 will hold in this case.

Theorem 26.7. *If r_i, $i = 1, \ldots, n_0$, are such that the system of inequalities*

$$Ax \le b, \qquad x \ge 0, \qquad \|x^i\|_i < r_i, \qquad i = 1, \ldots, n_0$$

is compatible, and problem (26.6) is solvable, then problem ($26.6^\#$) is solvable also. Moreover, their optimal values coincide. Conversely, solvability of problem ($26.6^\#$) yields solvability of problem (26.6); and their optimal values coincide.

Theorem 26.8. *Suppose the system of restrictions in (26.6) is compatible (this provides boundedness of optimal value of problem ($26.6^\#$) by Theorem 25.1) and the lower bound in ($26.6^\#$) is accessible (i.e. ($26.6^\#$) is solvable). The problem (26.6) is solvable also, and the optimal values of problems (26.6) and ($26.6^\#$) coincide.*

Note that if $B_0 = \varnothing$, then problem (26.6) is solvable if only its restriction system is compatible. This follows from the fact that in this case $x = [x^1, \ldots, x^{n_0}]^T$ and, therefore, inequalities $\|x_i\|_i \le r_i$, $i = 1, \ldots, n_0$, define the bounded set of solutions. In this situation problem ($26.6^\#$) can be written as follows:

$$\min\left\{(b, u) + \sum_{i=1}^{n_0} r_i \|(c^i - B_i^T u)^+\|_i^* \ \bigg| \ u \ge 0\right\}.$$

If, in addition, $n_0 = 1$ (therefore, $B_1 = A$), then problems (26.6) and ($26.6^\#$) attain the forms

$$\max\{(c, x) \mid Ax \le b, \ x \ge 0, \ \|x\|_q \le r_0\}, \qquad (26.7)$$

$$\min\{(b, u) + r_0\|(c - A^T u)^+\|_q^* \mid u \geq 0\} \qquad (26.7^\#)$$

where $r_0 > 0$, $\|\cdot\|_q$ – is an arbitrary monotone norm in the space $\mathbb{E}_n$. For this pair of problems the analog of Theorem 26.6 holds also.

Theorem 26.9. *If $r_0 > \min\{\|x\|_q \mid x \geq 0, Ax \leq b\}$ in (26.7), then each problem (26.7) and (26.7$^\#$) is solvable and their optimal values coincide.*

27.   THE DUALITY THEOREM FOR L-PROBLEMS

We recall that if in (25.2) the norms $\|\cdot\|_j$ and $\|\cdot\|_i$ have the types $\|\cdot\|_0$ or $\|\cdot\|_1$, this problem is defined as *l–problem*. If (25.2) is a l–problem, then, from the relations $\|\cdot\|_0^* = \|\cdot\|_1$, $\|\cdot\|_1^* = \|\cdot\|_0$ it follows that (25.2$^\#$) is a l–problem also (and conversely).

Suppose now that (25.2) is a l–problem. Then, this problem (and (25.2$^\#$) also) is a convex piecewise linear program and can be rewritten in the form of *LP* problem. For l–problems we do not discuss the problems of accessibility of lower and upper bounds in (25.2) and (25.2$^\#$) just as the problem of regularity conditions which provide the possibility of application of the Kun-Takker theorem for proving the duality theorem (see Remark 25.9).

Theorem 27.1. *If (25.2) is the l–problem then its solvability yields solvability of (25.2$^\#$) (and conversely) and coincidence their optimal values.*

If the admissible sets $M(r)$ and $M^\#(R)$ of l–problems (25.2) and (25.2$^\#$) are not empty, then these problems are solvable, and then their optimal values coincide.

We shall clarify the proof of the second part of the theorem. By Theorem 25.1 the optimal values of problems (25.2) and (25.2$^\#$) are finite. As for the l–problem the upper bound in (25.2) is accessible, then (25.2) is solvable and the theorem holds.

We consider now one particular case of l–problem; namely, problem (26.5) for $\|\cdot\|_p = \|\cdot\|_1$. Then there arise the following pair of mutually dual problems:

$$\max\left\{(c, x) - R_0 \sum_{j=1}^{m} l_j^+(x) \;\Big|\; x \geq 0\right\}, \qquad (27.1)$$

$$\min\{(b, u) \mid A^T u \geq c, \ 0 \leq u \leq \bar{R}\} \tag{27.1\#}$$

where $R_0 > 0$, $\bar{R} = [R_0, \ldots, R_0]^T \in \mathbb{E}_m$, $[l_1^+(x), \ldots, l_m^+(x)]^T = (Ax - b)^+$. These problems correspond to the inconsistent problem (25.1) and are connected by regular duality.

The two following dual problems

$$\max\left\{(c, x) - \sum_{j>k} R_j l_j^+(x) \ \Big| \ A_0 x \leq b^0, \ x \geq 0\right\}, \tag{27.2}$$

$$\min\{(b, u) \mid A^T u \geq c, \ u \geq 0, \ u_j \leq R_j, \ \forall j > k\} \tag{27.2\#}$$

are also interesting. They are obtained from (26.3) and (26.3#) for $A_0 x - b^0 = [l_1(x), \ldots, l_k(x)]^T$, $A_j x - b^j = l_j(x)$, $\forall j > k$. The pair (27.1) and (27.1#) is the particular case of (27.2) and (27.2#).

In conclusion, we shall write down the problems C and $C^\#$ for another standard forms of the initial LP problem, namely

$$L_1 : \max\{(c, x) \mid Ax \leq b\},$$
$$L_2 : \min\{(c, x) \mid Ax = b, \ x \geq 0\}.$$

The dual problems to L_1 and L_2 will be written as follows:

$$L_1^* : \min\{(b, u) \mid A^T u = c, \ u \geq 0\},$$
$$L_2^* : \max\{(b, u) \mid A^T u \leq c\}.$$

The corresponding problems C and $C^\#$ will be as follows:

$$C_1 : \quad \sup\left\{(c, x) - \sum_{j=1}^{m_0} R_j \|(A_j x - b^j)^+\|_j \ \Big| \ A_0 x \leq b^0, \ \|x^i\|_i \leq r_i, \right.$$
$$\left. i = 1, \ldots, n_0\right\};$$

$$C_1^\# : \quad \inf\left\{(b, u) + \sum_{i=1}^{n_0} r_i \|c^i - B_i^T u\|_i^* \ \Big| \ B_0^T u = c^0, \ u \geq 0, \ \|u^j\|_j^* \leq R_j, \right.$$
$$\left. j = 1, \ldots, m_0\right\};$$

$$C_2: \quad \inf\left\{(c,x) + \sum_{j=1}^{m_0} R_j \|A_j x - b^j\|_j \;\middle|\; A_0 x = b^0, \; x \geq 0, \; \|x^i\|_i \leq r_i,\right.$$
$$\left. i = 1, \ldots, n_0\right\};$$

$$C_2^\#: \quad \sup\left\{(b,u) - \sum_{i=1}^{n_0} r_i \|(c^i - B_i^T u)^+\|_i^* \;\middle|\; B_0^T u \leq c^0, \; \|u^j\|_j^* \leq R_j,\right.$$
$$\left. j = 1, \ldots, m_0\right\}.$$

The formulation of Duality Theorems 25.2 and 27.1 for initial problems L_1 and L_2 can be written easily. We have given form of C_1 and $C_1^\#$, C_2 and $C_2^\#$ in order to give the possibility to reproduce the scheme 2 for arbitrary representation of the initial problem L.

Remark 27.1. We have considered the situation when the problems C and $C^\#$ are formed with the use of norms $\|\cdot\|_j$ and $\|\cdot\|_i$ of piecewise linear type; more exactly these are the norms

$$\|z\|_0 = \max_i |z_i| \qquad \text{and} \qquad \|z\|_1 = \sum_i |z_i|.$$

Indeed, all, that was formulated in Section 27, hold and for arbitrary piecewise linear norms $\|z\|$, which are given by the formula

$$\|z\| = \max_k |(a_k, z)|.$$

In this case, among the vectors $\{a_k\}$ there are n linearly independent, where n is the dimension of the space of variable x. If l–problem (25.2) is the problem where all the norms are piecewise linear and monotone (together with the conjugate norms), then Theorem 27.1 will hold true. In this paragraph we have considered the norms $\|\cdot\|_0$ and $\|\cdot\|_1$ for simplicity of presentation.

Chapter 4.

Problems of successive linear programming and duality

In Section 21.2 certain problems related to problems of successive (lexicographic) programming were considered. In this chapter, we consider similar problems, however, the main goal is to develop *symmetric duality* for these problems. *Symmetric duality* for successive programming is the duality applied to such initial problem whose dual object is a problem of successive programming also. In this case, the both problems are connected by informative mathematical relations.

We shall clarify now the sense of equivalent use of terms successive and lexicographic optimization. Let $z = [z_1, \ldots, z_k]^T \in \mathbb{E}_k$. We define *the lexicographic order* (lex-order) $\overset{(lex)}{\leq}$ in $\mathbb{E}_k$ which corresponds to the order $p = (i_1, \ldots, i_k)$ of indices of coordinates of the vector z as follows: $z \overset{(lex)}{\leq} z'$ is the first pair satisfies the inequality $z_{i_t} \leq z'_{i_t}$ with respect to the index t of distinct coordinates.

Coincidence of all coordinates corresponds to the equality $z = z'$.

Otherwise, $z \overset{(lex)}{\leq} z'$ (with respect to the order p) if and only if $z_{i_1} < z'_{i_1}$; if $z_{i_1} = z'_{i_1}$ then $z_{i_2} \leq z'_{i_2}$ and so on.

The order introduced above defines extremal elements in a certain set $Z \subset \mathbb{E}_k$. Namely, $\tilde{z} \in Z$ is *lex-maximal* in Z if

$$\bar{z} \overset{(lex)}{\geq} \tilde{z}, \qquad \bar{z} \in Z, \qquad \bar{z} = \tilde{z}.$$

The lex-minimal element may be defined similarly. The notion of lex-extremum (lex-minimum or lex-maximum) is connected with successive programming in the following sense. In the situation of the lex-maximal element $\tilde{z}$ we find the optimal vector of the final problem:

$$\max\{z_{i_1} \mid z \in Z\}, \tag{$*_1$}$$

$$\max\{z_{i_2} \mid z \in \text{Arg}\,(*_1)\}, \tag{$*_2$}$$

$$\cdots\cdots\cdots\cdots\cdots\cdots \qquad \vdots$$

$$\boxed{\max\{z_{i_k} \mid z \in \text{Arg}\,(*_{k-1})\}.} \tag{$*_k$}$$

Lexicographic optimization in whole, for example, for the problem

$$\max_p \left\{ \begin{bmatrix} f_1(x) \\ \vdots \\ f_k(x) \end{bmatrix} \,\middle|\, x \in M \right\} \tag{$*$}$$

is a problem of determining the element $\tilde{z}$ which is lex-maximal (with respect to a certain order $p = (i_1, \ldots, i_k)$ of the numbers of functions $\{f_i(x)\}_1^k$) in the set

$$Z = \left\{ \begin{bmatrix} z_1 \\ \vdots \\ z_k \end{bmatrix} \,\middle|\, \begin{bmatrix} z_1 \\ \vdots \\ z_k \end{bmatrix} \leq \begin{bmatrix} f_1(x) \\ \vdots \\ f_k(x) \end{bmatrix}, \; x \in M \right\}.$$

The sequence of problems $(*_1)$–$(*_k)$ for $(*)$ is transformed into the sequence of problems

$$\max\{f_{i_1}(x) \mid x \in M\}, \tag{$**_1$}$$

$$\max\{f_{i_2}(x) \mid x \in \text{Arg}\,(**_1)\}, \tag{$**_2$}$$

$$\cdots\cdots\cdots\cdots\cdots\cdots\cdots \qquad \vdots$$

$$\boxed{\max\{f_{i_k}(x) \mid x \in \text{Arg}\,(**_{k-1})\}.} \tag{$**_k$}$$

If $\tilde{x} \in \text{Arg}\,(**_k)$, then $\tilde{z}[f_1(\tilde{x}), \ldots, f_k(\tilde{x})]^T$ is a lex-maximal element from Z. The problems considered below will be clarified for linear settings.

28.   THE SCHEME OF DUALITY FORMATION IN LINEAR SUCCESSIVE PROGRAMMING

Let

$$L_{lex} : \max_{p} \left\{ \begin{bmatrix} c_0^T x \\ C^T x \end{bmatrix} \Bigg| Ax \le b, \ x \ge 0 \right\} \qquad \text{where} \quad C^T x = \begin{bmatrix} c_1^T x \\ \vdots \\ c_k^T x \end{bmatrix}$$

be a problem of lexicographic programming for an order $p = (k, \dots, 1, 0)$ of criterions $\{c_j^T x\}_0^k$. Here $C^T x = \begin{bmatrix} c_1^T x \\ \vdots \\ c_k^T x \end{bmatrix}$. Symmetrization of duality for this problem can be done using the parameter representation of the vector b. We set

$$b = b_0 + \sum_{i=1}^{l} r_i b_i = b_0 + Br, \qquad r_i \ge 0.$$

Here r_i are certain parameters. Then the initial problem can be written as follows:

$$L_{lex} : \max_{p} \left\{ \begin{bmatrix} c_0^T x \\ C^T x \end{bmatrix} \Bigg| Ax \le b_0 + Br, \ x \ge 0 \right\}. \qquad (28.1)$$

The dual problem following symmetry sense is written as follows:

$$L_q^* : \min_{q} \left\{ \begin{bmatrix} b_0^T u \\ B^T u \end{bmatrix} \Bigg| A^T u \ge c_0 + CR, \ u \ge 0 \right\}. \qquad (28.2)$$

Here $\{R_j \ge 0\}_1^k$ are parameters, q is the order of the functions $\{b_j^T u\}_0^l$. We take $q = (l, \dots, 1, 0)$. We shall call problem (28.2) lexicographic-dual to problem (28.1). The orders p and q in direct and dual problems respectively may be arbitrary. For definiteness we suppose $p = (k, \dots, 1, 0)$ and $q = (l, \dots, 1, 0)$. This, in particular, allows us to write our relations more compactly.

The exact sense of the problem of successive programming was given in Section 21; however, we shall formalize if in application to problems L_p and L_q^*.

First, we write the two groups of problems:

$$\max\{c_k^T x \mid Ax \leq b_0 + Br, \ x \geq 0\}, \qquad (28.3_k)$$

$$\max\{c_{k-1}^T x \mid x \in \operatorname{Arg}(28.3_k)\}, \qquad (28.3_{k-1})$$

....

$$\boxed{\max\{c_0^T x \mid x \in \operatorname{Arg}(28.3_1)\}} \qquad (28.3_0)$$

– is the first group. The second group is as follows:

$$\min\{b_l^T u \mid A^T u \geq c_0 + CR, \ u \geq 0\}, \qquad (28.4_l)$$

$$\min\{b_{l-1}^T u \mid u \in \operatorname{Arg}(28.4_l)\}, \qquad (28.4_{l-1})$$

....

$$\boxed{\min\{b_0^T u \mid u \in \operatorname{Arg}(28.4_1)\}} \qquad (28.4_0)$$

The problems L_p and L_q^* are exactly the final problems (28.3_0) and (28.4_0) from these groups.

The problems L_p and L_q^* may be more generally widening the notions of their optimal sets. Namely, we set:

$$\operatorname{Arg}_+ L_p := \{[\bar{x}_0, \ldots, \bar{x}_k] \mid \bar{x}_t \in \operatorname{Arg}(28.3)_t, \ t = 0, \ldots, k\},$$

$$\operatorname{Arg}_+ L_q^* := \{[\bar{u}_0, \ldots, \bar{u}_l] \mid \bar{u}_s \in \operatorname{Arg}(28.4)_s, \ s = 0, \ldots, l\}.$$

We shall use below this widening of solutions of problems L_p and L_q^*.

So, we take problem (28.1) as the initial problem of lex-optimization; the dual problem is (28.2) which is a lex-optimization problem also. Note that some parameters of these problems are not connected. These problems are r, R, p and q. Duality between L_p and L_q^* is a somewhat external. Their mathematical connection is revealed via problems which scalarize them in a cross way, namely:

$$L_{p,scal}: \ \max\{c_0^T x + (C^T x, R) \mid Ax \leq b_0 + Br, \ x \geq 0\}, \qquad (28.5)$$

$$L_{q,scal}^*: \ \min\{b_0^T u + (B^T u, r) \mid A^T u \geq c_0 + CR, \ u \geq 0\}. \qquad (28.5^*)$$

These problems are problems of linear programming dual in classical sense.

The duality theorem shows both the connection between L_p and L_q^* and between $L_{p,scal}$ and $L_{q,scal}^*$. This connection can be seen from the scheme

$$L_p \xrightarrow{\text{Arg} \atop \sim} L_{p,\,scal}$$

$$(lex) \Big\updownarrow \qquad \qquad \Big\updownarrow (*)$$

$$L_q^* \xrightarrow{\text{Arg} \atop \sim} L_{q,\,scal}^*$$

Scheme 3

In this scheme the symbol $\xrightarrow{\text{Arg} \atop \sim}$ denotes the transfer from one problem to the other problem in the case when their optimal sets coincide. The realization of this scheme is the sense of the duality theorem.

29. SOLVABILITY CONDITIONS FOR LEXICOGRAPHIC OPTIMIZATION PROBLEMS

We consider the problem L_{lex} from Section 28

$$\max_p \left\{ \begin{bmatrix} c_0^T x \\ C^T x \end{bmatrix} \;\middle|\; Ax \le b, \; x \ge 0 \right\} \tag{29.1}$$

when $p = (k, \dots, 1, 0)$. The conditions of its solvability can be easily derived from the conditions of solvability for LP problems. Namely, $M \ne \varnothing$ & $M^* \ne \varnothing$, where M is the admissible set for the dual problem (Theorem 15.4). If the problems L and L^* have the form (22.1), (22.2), then $M = \{x \ge 0 \mid Ax \le b\}$; $M^* = \{u \ge 0 \mid A^T u \ge c\}$. We write the restrictions for the problem L and L^* in the form

$$\begin{array}{ll} Ax + v = b; & A^T u - w = c; \\ [x, v] \ge 0; & [u, w] \ge 0; \end{array} \tag{29.2}$$

where $v \in \mathbb{E}_m$, $w \in \mathbb{E}_n$. Suppose a_j is the j–row of the matrix A; h_i is its i–column; $\{e_j\}_1^m$ are unit vectors of the space $\mathbb{E}_m$; $\{e_i^*\}_1^n$ are unit vectors of the space $\mathbb{E}_n$. In this case, systems (29.2) can be written as follows:

$$\begin{cases} \sum\limits_{i=1}^n x_i h_i + \sum\limits_{j=1}^m v_j e_j = b, \\ x_i \ge 0, \quad v_j \ge 0, \\ i = 1, \dots, n, \\ j = 1, \dots, m; \end{cases} \qquad \begin{cases} \sum\limits_{j=1}^m u_j a_j^T - \sum\limits_{i=1}^n w_i e_i^* = c, \\ u_j \ge 0, \quad w_i \ge 0, \\ j = 1, \dots, m, \\ i = 1, \dots, n. \end{cases} \tag{29.3}$$

Solvability of systems (29.3) denotes that

$$b \in \text{cone}\{h_i, e_j\}_{1,1}^{n,m}, \qquad c \in \text{cone}\{a_j^T, -e_i^*\}_{1,1}^{m,n}. \tag{29.4}$$

This for of solvability of the problem L (therefore, of the problem L^* also) reduces the compact form for solvability condition of problem (29.1).

Theorem 29.1. *If the restriction system for problem (29.1) is solvable, then problem (29.1) is solvable if and only if the following relations*

$$\begin{cases} c_k \in \text{cone}\{a_j^T, -e_i^*\}_{1,1}^{m,n}, \\ c_{k-1} \in \text{cone}\{-c_k, J\}, \\ \quad\cdots\cdots\cdots\cdots\cdots\cdots \\ c_0 \in \text{cone}\{-c_1, \ldots, -c_k, J\}, \end{cases} \tag{29.5}$$

hold. Here $J = \{a_j^T, -e_i^*\}_{1,1}^{m,n}$.

Remark 29.1. Relations (29.5) do not depend of the vector b of right-hand sides of restrictions of problem (29.1). This vector defines only the condition $M \neq \varnothing$, which is equivalent to realization of the left inclusion in (29.4).

Remark 29.2. Conditions (29.5) were first used in Section 21 in application to problem (21.12) as its solvability conditions. We have repeated these conditions in order to show the symmetric conditions of solvability of the problem lexicographically dual to (29.1).

We write now the lex-problem whose form is dual to (29.1)

$$\min\left\{ \begin{bmatrix} b_0^T u \\ B^T u \end{bmatrix} \;\middle|\; A^T u \geq c, \; u \geq 0 \right\}. \tag{29.6}$$

Problems (29.1) and (29.6) are not coordinated yet by the information. Indeed, we shall be interested in these problems for $b = b_0 + BR$, $c = c_0 + Cr$.

Theorem 29.2. *If the restriction system for problem (29.6) is solvable, then problem (29.6) is solvable if and only if the relations*

$$\begin{cases} b_l \in \text{cone}\{h_i, e_j\}_{1,1}^{n,m}, \\ b_{l-1} \in \text{cone}\{-b_l, I\}, \\ \quad\cdots\cdots\cdots\cdots\cdots\cdots\cdots \\ b_0 \in \text{cone}\{-b_1, \ldots, -b_l, I\}, \end{cases} \tag{29.7}$$

hold. Here $I = \{h_i, -e_j\}_{1,1}^{n,m}$.

Remark 29.3. If the restriction systems of problems (29.1) and (29.6) are compatible simultaneously, we cannot guarantee solvability of these problems.

Nevertheless, we can formulate the following statement on simultaneous solvability of problems (29.1) and (29.6) for $c = c_0 + Cr$ and $b = b_0 + Br$.

Theorem 29.3. *Suppose, in problems (29.1) and (29.6) $c = c_0 + CR$, $b = b_0 + BR$ and $r \geq 0$, $R \geq 0$ are such that*

$$M(r) = \{x \geq 0 \mid Ax \leq b_0 + Br\} \neq \emptyset,$$
$$M^*(R) = \{u \geq 0 \mid A^T u \geq c_0 + CR\} \neq \emptyset.$$

Then, problems (29.1) and (29.6) are solvable if and only if (29.5) and (29.7) hold.

We shall deduce some conditions which will provide the theorem on lexicographic duality in various variants.

Condition A₁. The systems $\{Ax \leq b_s, x \geq 0\}_0^l$ and $\{A^T u \geq c_t, u \geq 0\}_0^k$ are compatible.

Condition A₂. The systems of vectors $\{a_j^T, -e_i^*\}_{1,1}^{m,n}$ and $\{h_i, e_j\}_{1,1}^{n,m}$ are multifold.

Condition A₃. Each of the systems 1) $Ax \leq b_0$, $x \geq 0$; 2) $A^T u \geq c_0$, $u \geq 0$; 3) $Br \geq 0$, $r > 0$; 4) $CR < 0$, $R > 0$ is compatible (the compatibility of systems 3 and 4 is considered relative to the parameters r and R).

Now, we shall explain the conditions A_1–A_3. The Condition A_1 guarantees compatibility of the systems $Ax \leq b_0 + Br$, $x \geq 0$ and $A^T u \geq c_0 + CR$, $u \geq 0$ for each $r \geq 0$ and $R \geq 0$.

Really if $\bar{x}_s$ is a solution of the system $Ax \leq b_s$, $x \geq 0$ and $\bar{u}_t$ is a solution of the system $A^T u \geq c_t$, $u \geq 0$, then $\bar{x} = \bar{x}_0 + \sum_{s=1}^{k} R_s \bar{x}_s$ and $\bar{u} = \bar{u}_0 + \sum_{t=1}^{l} r_t \bar{u}_t$ are the solutions of these systems (with the right-hand sides $b_0 + Br$ and $c_0 + CR$) respectively.

The Condition A_1 can be used in Theorems 29.2 and 29.3 in the situation when, in particular, for the formulation of theorem of lexicographic duality, we need to choose certain values of parameter r and R.

The Condition A_2 guarantees the boundedness of polyhedrons $M = \{x \geq 0 \mid Ax \leq b\}$ and $M^* = \{u \geq 0 \mid A^T u \geq c\}$ if $M \neq \varnothing$, $M^* \neq \varnothing$ (Theorems 3.1 and 3.2). If the polyhedron is bounded, then each lexicographic problem is solvable. Therefore, if we combine the conditions A_1 and A_2 we obtain solvability of the problems L_p and L_q^* for each values of parameters $r \geq 0$ and $R \geq 0$ and orders p and q.

The Condition A_3 is weaking condition of solvability of the systems $Ax \leq b_0 + Br$, $x \geq 0$ and $A^T u \geq c_0 + CR$, $u \geq 0$ for each $r > 0$, $R > 0$.

30.   THE DUALITY THEOREM

As it was noted in the end of Section 28, the contents of the duality theorem for the lexicographic problems L_p and L_q^* is to realize the Scheme 3 from Section 28. We formulate first the statement similar to Theorem 21.5 in application to the problems L_p and $L_{p,scal}$ (as well as to L_q^* and $L_{q,scal}^*$).

Theorem 30.1. *1. Let the problem L_p, i.e. (28.1), be solvable for a certain $r \geq 0$. Then there exists a nonempty domain of values of the parameter $R \geq 0$ so that*

$$\operatorname{Arg} L_p = \operatorname{Arg} L_{p,scal} \tag{30.1}$$

where the choice of parameters R is independent of r.

2. Let the problem L_q^*, i.e. (28.2), be solvable for a certain $R \geq 0$. Then there exists a nonempty set of values of the parameter $r \geq 0$, so that

$$\operatorname{Arg} L_q^* = \operatorname{Arg} L_{q,scal}^* \tag{30.2}$$

where the choice of parameters r is independent of R.

As it follows from Theorem 21.5, these statements hold true.

Theorem 30.2 [duality]. *Suppose the conditions* A_1, *(29.5) and (29.7) hold for the problems* L_p *and* L_q^* *(these are the conditions which provide solvability of problems* L_p *and* L_q^* *for each parameter* r *and* R *from the right-hand sides of their restriction systems). Then there exists such a nonempty domain of values of these parameters that the Scheme 3 is realized. The sense of this scheme is that it contains equalities (30.1), (30.2), and* $\operatorname{opt} L_{p,\,scal} = \operatorname{opt} L_{q,\,scal}^*$.

Proof. The proof of this theorem is anticipated by Theorem 30.1; more exactly, by Theorem 21.5. Really, the Condition A_1 provides (taking into account its clarifying) the condition $M(\bar{r}) \neq \varnothing$, $M^*(\bar{R}) \neq \varnothing$ for each $\bar{r} \geq 0$ and $\bar{R} \geq 0$. Conditions (29.5) and (29.7) provide solvability of the problems L_p and L_q^*. Independently of fixed $\bar{r}$ and $\bar{R}$, by Theorem 30.1, we can choose such $R \geq 0$ and $r \geq 0$ that relations (30.1) and (30.2) will hold for problems L_p, L_q^*, $L_{p,\,scal}$, and $L_{q,\,scal}^*$ with the parameters $\bar{r}$ and $\bar{R}$ standing in the right-hand sides of their restrictions. If we replace them by r and R that we find, then these relations will be the desired relations relatively Scheme 3. As the problems $L_{p,\,scal}$ and $L_{q,\,scal}^*$ are dual in classical sense, their optimal values coincide (Theorem 15.2), i.e. $\operatorname{opt} L_{p,\,scal} = \operatorname{opt} L_{q,\,scal}^*$. □

Remark 30.1. In formulation of Theorem 30.2 we have used the conditions A_1, (29.5) and (29.7). The two last conditions can be replaced by A_2.

The following variant of conditions which guarantee the realization of Scheme 3 is possible also. Namely, in order that the restriction systems $Ax \leq b_0 + Br$, $x \geq 0$ and $A^T u \geq c_0 + CR$, $u \geq 0$ be solvable, we introduce the condition A_3. In order that the lexicographic problems L_p and L_q^* be solvable we introduce the condition A_2. This means that we introduce the conditions A_2 and A_3 in the whole.

Remark 30.2. In Section 28 we have introduced the notion of widen optimal set $\operatorname{Arg}_+(\cdot)$ for the lexicographic problems L_p and L_q^* (see 28.3_t) and (28.4_s). In this connection, Theorems 30.1 and 30.2 holds if we replace in relations (30.1) and (30.2) $\operatorname{Arg}(\cdot)$ by $\operatorname{Arg}_+(\cdot)$.

31. REDUCTION OF LEXICOGRAPHIC OPTIMIZATION PROBLEMS TO SYSTEMS OF LINEAR INEQUALITIES

As it follows from the duality theorem for the LP problems (15.1) and (15.1*), identification of their optimal sets is realized by the system of linear inequalities (15.2). In this case, solution of the problems L and L^* is reduced to solving a certain system of linear inequalities. Such reduction is possible for lexicographic problems of optimization.

31.1. A scheme of reduction for the two-step optimization problem

We consider here a particular case of problem (29.1) for $k = 1$, i.e.,

$$\max_p \left\{ \left[\begin{array}{c} (c_0, x) \\ (c_1, x) \end{array} \right] \;\middle|\; Ax \le b, \; x \ge 0 \right\} \tag{31.1}$$

$p = (0, 1)$. As it was above noted, the problem

$$\max\{(c_1, x) \mid Ax \le b, \; x \ge 0\} \tag{31.2}$$

is reduced to the system

$$\left\{ \begin{array}{l} Ax \le b, \quad x \ge 0; \quad A^T u \ge c_1, \quad u \ge 0; \\ \qquad (c_1, x) \ge (b, u). \end{array} \right. \tag{31.3}$$

Therefore, lex-problem (31.1) is the following problem of linear programming

$$\max\{(c_0, x) \mid (31.3)\,\} \tag{31.4}$$

or

$$\max\{(\bar{c}_0, z \mid \bar{A}z \le \bar{b}, \; z \ge 0\}, \tag{31.5}$$

where

$$\bar{c}_0 = \left[\begin{array}{c} c_0 \\ 0 \end{array} \right], \qquad z = \left[\begin{array}{c} x \\ u \end{array} \right], \qquad \bar{b} = \left[\begin{array}{c} b \\ -c_1 \\ 0 \end{array} \right], \qquad \bar{A} = \left[\begin{array}{ccc} A & & 0 \\ 0 & & -A^T \\ -c_1^T & & b^T \end{array} \right]$$

The following problem will be dual to (31.5):

$$\min\{(\bar{b}, \delta) \mid \bar{A}^T \delta \ge \bar{c}_0, \; \delta \ge 0\}, \tag{31.5*}$$

$$\delta = \begin{bmatrix} v \\ w \\ v_0 \end{bmatrix} \in \mathbb{E}_{m+n+1} \qquad (v \in \mathbb{E}_m, \quad w \in \mathbb{E}_n, \quad v_0 \in \mathbb{R} = \mathbb{E}_1).$$

The problems (31.5) and (31.5*) are reduced (following the above scheme) to the system

$$\bar{A}z \le \bar{b}, \quad \bar{A}^T\delta \ge \bar{c}_0, \quad z \ge 0, \quad \delta \ge 0,$$
$$(c_0, z) \ge (\bar{b}, \delta)$$

or, in the expanded form

$$\left\{ \begin{array}{c} Ax \le b, \quad x \ge 0; \quad A^T u \ge c_1, \quad u \ge 0; \\ (c_1, x) \ge (b, u); \\ A^T v - v_0 c_1 \ge c_0; \quad -Aw + v_0 b \ge 0, \\ v \ge 0, \quad w \ge 0, \quad v_0 \ge 0; \\ (c_0, x) \ge (b, v) - (c_1, w). \end{array} \right. \qquad (31.6)$$

In whole, the variable vector in system (31.6) in $[x, u, v, w, v_0]^T \in \mathbb{E}_{2n+2m+1}$. The result is as follows: problem (31.1) is reduced to system (31.6). If $[\bar{x}, \bar{u}, \bar{v}, \bar{w}, \bar{v}_0]^T$ is its solution, then $\bar{x} \in \text{Arg}\,(31.1)$ (simultaneously $\bar{x} \in \text{Arg}\,(31.2)$); $[\bar{x}, \bar{u}] \in \text{Arg}\,(31.5)$, $[\bar{v}, \bar{w}, \bar{v}_0] \in \text{Arg}\,(31.5^*)$.

31.2. General theorem on reducibility

Theorem 31.1. *Solution of solvable problem (29.1) is reduced to solution of a certain system of linear inequalities formed constructively. Moreover, if $\widetilde{M}$ is the algebraic component with respect to x of the solution set of this system, then $\widetilde{M} = \text{Arg}\,(29.1)$.*

Proof. The proof of this theorem is the subsequent application of the reduction scheme from Section 31.1. $\square$

32. LEXICOGRAPHIC DUALITY FOR IMPROPER LP PROBLEMS—A SPECIAL CASE

We shall consider below the *LP* problem in a usual setting. However, we shall take into account the order (with respect to importance) of restrictions both in the direct problem L and the dual problem L^*. Such setting defines a unique choice of maximally compatible subsystems (MCS) in their restriction systems. Suppose, we have

$$L: \max\{(c, x) \mid Ax \leq b, \ x \geq 0\},$$

$$L^*: \min\{(b, u) \mid A^T u \geq c, \ u \geq 0\}.$$

The order of inequalities in restrictions we define by permulations $p = (j_m, \ldots, j_1)$, $q = (i_n, \ldots, i_1)$. This denotes that j_{l+1}-inequality in the system $Ax \leq b$ in more important then j_l-inequality ; and i_{t+1}-inequality in the system $A^T u \geq c$ is more important than i_l-inequality; $l = 1, \ldots, m - 1$, $t = 1, \ldots, n - 1$.

For definiteness, we shall assume that $p = (1, \ldots, m)$, $q = (1, \ldots, n)$. MSC is formed by accumulating the number of inequalities beginning with the last one (inclusion of the condition $x \geq 0$ into MSC is assumed). In such situation the skips are possible. These are the situations when the inclusion of recurrent inequality leads to incompatibility. However, the inclusion of the next inequality gives the compatible subsystem. We may have several skips. This situation may be considered as revision of the initial order of restrictions (by their importance).

Without loss of generality we assume that skips are absent. The selected MSC in this case will be as follows:

$$l_j(x) := (a_j, x) - b_j \leq 0, \qquad x \geq 0, \qquad j = k + 1, \ldots, m; \qquad (32.1)$$

$$-h_i(x) := (h_i, u) - c_i \geq 0, \qquad u \geq 0, \qquad i = s + 1, \ldots, n. \qquad (32.2)$$

Here

$$Ax - b \leq 0 \quad \sim \quad l_j(x) \leq 0, \qquad j = 1, \ldots, m;$$

$$A^T u - c \geq 0 \quad \sim \quad -h_i(u) \geq 0, \qquad i = 1, \ldots, n.$$

The set of solutions of (32.1) and (32.2) we denote as M_0 and M_0^*.

For inequalities which do not enter systems (32.1) and (32.2) we find the minimal residuals on the basis of the principle of lexicographic minimum. Now, we formulate the problems

$$L_p(r): \quad \max_p \left\{ \begin{bmatrix} (c, x) \\ -l_1^+(x) \\ \cdots \\ -l_k^+(x) \end{bmatrix} \ \middle| \ x \geq 0, \ x \in M_0, \ x^1 \leq r \right\},$$

$$L_q^\#(R): \quad \min_q \left\{ \begin{bmatrix} (b,u) \\ h_1^+(u) \\ \cdots \\ h_s^+(u) \end{bmatrix} \;\middle|\; u \geq 0, \; u \in M_0^*, \; u^1 \leq R \right\},$$

where $p = (0,1,\ldots,k)$, $q = (0,1,\ldots,s)$. The zero indices in these permutations correspond to the goal functions (c,x) and (b,u); x^1 and u^1 are the components of the vectors x and u correspondent to partition of the system $Ax \leq b$ onto the subsystems $l_j(x) \leq 0$, $j = 1,\ldots,k$, and (32.1). The system $-h_i(u) \geq 0$ is decomposed onto the subsystems $-h_i(u) \geq 0$, $i = 1,\ldots,s$, and (32.2). We can depict this situation as follows:

Note that entrance of the restrictions $x^1 \leq r$ and $u^1 \leq R$ in the problems $L_p(r)$ and $L_q^\#(R)$ is essential.

Scalarization of the problem L_p by means of the vector R and the problem $L_q^\#$ by means of the vector r leads to the pair of problems

$$L_p(r,R): \quad \max \left\{ (c,x) - \sum_{j=1}^k R_j l_j^+(x) \;\middle|\; x \in M_0, \; x_1 \leq r \right\},$$

$$L_q^\#(R,r): \quad \min \left\{ (b,u) + \sum_{i=1}^l r_i h_i^+(u) \;\middle|\; u \in M_0^*, \; u_1 \leq R \right\}.$$

As M_0 and M_0^* are defined by correspondent MSC, then $l_j(x) \geq 0$, $\forall x \in M_0$, $j = 1,\ldots,k$ and $h_i(u) \geq 0$, $\forall u \in M_0^*$, $i = 1,\ldots,l$. This allows us to replace $l_j^+(x)$ and $h_i^+(u)$ by $l_j(x)$ and $h_i(u)$ for the marked indices j and i.

The problems $L_p(r,R)$ and $L_q^\#(R,r)$ are connected by the following statement (Section 25, Theorem 25.2).

If R and r are such that

$$M_0 \cap \{x \mid x^1 \le r\} \ne \varnothing, \qquad M_0^* \cap \{u \mid u^1 \le R\} \ne \varnothing$$

then these problems are solvable and their optimal values coincide.

The below theorem (taking into account the above statement) shows the connection between the problems L_p, $L_q^\#$, $L_p(r, R)$, and $L_q^\#(R, r)$.

Theorem 32.1. *There exists a nonempty domain of the parameters $r \ge 0$ and $R \ge 0$ defined constructively so that*

$$\mathrm{Arg}\, L_p(r) = \mathrm{Arg}\, L_p(r, R) \ne \varnothing,$$

$$\mathrm{Arg}\, L_q^\#(R) = \mathrm{Arg}\, L_q^\#(R, r) \ne \varnothing,$$

$$\mathrm{opt}\, L_p(r, R) = \mathrm{opt}\, L_q^\#(R, r).$$

Remark 32.1. If the problem L is solvable (then L^* is solvable also), then the problems $L_p(r)$ and $L_p(r, R)$ coincide with L and the problems $L_q^\#(R)$ and $L_q^\#(R, r)$ coincides with L^*. So, we have formulated Theorem 32.1 specially for unsovable problems of linear programming, where the restrictions are ordered (with the sense of order-importance).

33.   DUALITY FOR IMPROPER LP PROBLEMS IN LEXICOGRAPHIC INTERPRETATION

We suppose here that P and $P^\#$ are the problems C and $C^\#$ from Section 25.

The problem on approximate sense of problems P and $P^\#$ relatively L and L^* is of great interest. We shall clarify this sense using the terms of lexicographic optimization.

We shall assume that the systems of restrictions L and L^* are separated onto the subsystems $A_j x \le b^j$, $j = 1, \ldots, m_0$, and $B_i^T u \ge c^i$, $i = 1, \ldots, n_0$. The inequalities of each subsystem are not ordered; only the subsystems are ordered; for example, in correspondence with the orders $(0, m_0, \ldots, 1)$ and $(0, n_0, \ldots, 1)$. This order has the following sense. The restrictions $A_0 x \le b^0$, $x \ge 0$ and $B_0^T u \ge c^0$, $u \ge 0$ are of *directive* character and they are compatible. The other restrictions are *facultative* (they may be incompatible).

Taking into account this order of subsystems, the functions

$$f_j(x) = \|(A_j x - b^j)^+\|_j, \qquad j = 1, \dots, m_0,$$
$$g_i(u) = \|(c^i - B_i^T u)^+\|_i^*, \qquad i = 1, \dots, n_0,$$

which are residual of the correspondent systems, must be minimized subsequently following the order $p = (m_0, \dots, 1)$ and $q = (n_0, \dots, 1)$. As it was noted in previous sections, this is equivalent to setting the two following problems of lexicographic optimization

$$P_p(r): \quad \max_p \left\{ \begin{bmatrix} -f_{m_0}(x) \\ \cdots \\ -f_1(x) \\ (c, x) \end{bmatrix} \, \middle| \, x \in M(r) \right\},$$

$$P_q^\#(R): \quad \min_q \left\{ \begin{bmatrix} g_{n_0}(u) \\ \cdots \\ g_1(u) \\ (b, u) \end{bmatrix} \, \middle| \, u \in M^\#(R) \right\},$$

where $M(r)$ and $M^\#(R)$ are the admissible sets for the problems P and $P^\#$ respectively. The symbols p and q define the order of residual with the condition that the functions (c, x) and (b, u) in these orders are situated in the end.

Theorem 33.1. *Suppose that the norms* $\{\|\cdot\|_j\}$, $\{\|\cdot\|_i\}$ $r \geq 0$ *and* $R \geq 0$ *are monotone together with the adjoint norms and they are piecewise linear.*

$$\operatorname{Arg} P = \operatorname{Arg} P_p(r) \neq \varnothing, \tag{33.1}$$

$$\operatorname{Arg} P^\# = \operatorname{Arg} P_q^\#(R) \neq \varnothing, \tag{33.2}$$

$$\operatorname{opt} P = \operatorname{opt} P^\#. \tag{33.3}$$

Proof. We shall give explanation to the proof Relation (33.3) holds by the basis duality theorem for the improper *LP* problems (Theorem 25.2). As for relations (33.1) and (33.2), they are justified similar as in Theorem 30.2. We must refine only one moment. When we define $R_{m_0} > \bar{u}_{m_0}$, where $\bar{u}_{m_0} \geq 0$ is the dual estimate of restriction $f_{m_0}(x) = \|(A_{m_0}x - b^{m_0})^+\|_{m_0} \leq \alpha_{m_0}$ ($:= \text{opt} \min\{f_{m_0}(x) \mid x \in M(r)\}$) in the problem

$$\min\{f_{m_0-1}(x) \mid x \in M(r), \; f_{m_0}(x) \leq \alpha_{m_0}\} \tag{33.4}$$

then existence of $\bar{u}_{m_0}$ will be provided by the regularity condition for the restriction system for problem (33.4). This condition holds since the norms entering the problems P and $P^\#$ are piecewise linear. In this case the inequality $f_{m_0}(x) \leq \alpha_{m_0}$ and the inequalities $\|x^i\|_i \leq r_i$, $i = 1, \ldots, n_0$ can be written in the form of the system of linear inequalities. This fact guarantees the conditions of regularity both in problem (33.4) and in all other problems which arise when proving the theorem similar to the scheme of proving Theorem 30.2.

We consider the particular case of the problems $P_p(r)$ and $P_q^\#(R)$ in the form

$$\max_p \left\{ \begin{bmatrix} -\|(Ax - b)^+\|_s \\ (c, x) \end{bmatrix} \;\middle|\; x \geq 0, \; \|x\|_t \leq r \right\}, \tag{33.5}$$

$$\min_q \left\{ \begin{bmatrix} \|(c - A^T u)^+\|_t^* \\ (b, u) \end{bmatrix} \;\middle|\; u \geq 0, \; \|u\|_s^* \leq R \right\}. \tag{33.5*}$$

Here p and q correspond to the order of the functions from the top to the bottom. For example, in (33.5), first, we minimize the residual $\|(Ax-b)^+\|_s$, and in the Arg of this problem we maximize the function (c, x).

We shall be interested in the case of the improper problem L of the third order:

$$\min_{x \geq 0} \|(Ax - b)^+\|_s = \alpha > 0, \qquad \min_{u \geq 0} \|(c - A^T u)^+\|_t^* = \beta > 0,$$

(i.e., the systems $Ax \leq b$, $x \geq 0$ and $A^T u \geq c$, $u \geq 0$ are incompatible). Here s and t in $\|\cdot\|_s$ and $\|\cdot\|_t$ are the marks of the norms.

The problems P and $P^\#$ correspondent to the situation in question we write in the form

$$\max\{(c, x) - R\alpha\|(Ax - b)^+\|_s \mid x \geq 0, \; \|x\|_t \leq \beta r\}, \tag{33.6}$$

$$\min\{(b, u) + r\beta\|(c - A^T u)^+\|_t^* \mid u \geq 0, \|u\|_s^* \leq \alpha R\}. \qquad (33.6^\#)$$

It is clear that Theorem 33.1 holds true for this pair of problems. We want to replace the residuals in these problems by their squares which is often winning from the calculating sense. So, parallel with (33.6) and (33.6#), we write the problems

$$\max\{(c, x) - R\|(Ax - b)^+\|_s^2 \mid x \geq 0, \|x\|_t \leq \beta r\}, \qquad (33.7)$$

$$\min\{(b, u) + r\|(c - A^T u)^+\|_t^{*2} \mid u \geq 0, \|u\|_s^* \leq \alpha R\}. \qquad (33.7^\#)$$

Thus, (33.7) and (33.7#) differ from (33.6) and (33.6#) by the degrees of residual norms. Becides, we took away the multipliers α and β before the coefficients. In mathematical (in particular, in LP) programming the penalizing of residuals of the first or the second orders is essentially different. In the case of $\|(Ax - b)^+\|_s$ we obtain equivalent reduce of the initial LP problem to the problem without restrictions. In the case of $\|(Ax - b)^+\|_s^2$ we obtain the asymptotic reducibility.

This effect of distinction, in the case we consider, disappears. The reason in the condition $\alpha > 0$, $\beta > 0$. We shall show below, for certain assumptions (which were written above), that the following duality scheme holds

$$
\begin{array}{ccccc}
 & \xrightarrow{\mathrm{Arg}\,\sim} & & \xrightarrow{\mathrm{Arg}\,\subset} & \\
(33.5) & & (33.6) & & (33.7) \\
\updownarrow & & (\#)\,\updownarrow & & (\#)\,\updownarrow \\
 & \xrightarrow{\mathrm{Arg}\,\sim} & & \xrightarrow{\mathrm{Arg}\,\subset} & \\
(33.5^*) & & (33.6^\#) & & (33.7^\#)
\end{array}
$$

Scheme 4

Here $\xrightarrow{\mathrm{Arg}\,\subset}$ denotes $\mathrm{Arg}\,(33.6) \subset \mathrm{Arg}\,(33.7)$, $\mathrm{Arg}\,(33.6^\#) \subset \mathrm{Arg}\,(33.7^\#)$. Note, that by the duality theorem for improper LP problems, we have $\mathrm{opt}\,(33.6) = \mathrm{opt}\,(33.6^\#)$. $\square$

Theorem 33.2. *Suppose the norms $\|\cdot\|_s$ and $\|\cdot\|_t$ are monotone (together with the adjoint norms) and piecewise linear. Then there exists a nonempty set of the parameters $r > 0$, $R > 0$, which are constructively determined, such that the scheme above is realized.*

Proof. The left-hand side of the scheme holds by Theorem 33.1. We must justify now the inclusion Arg (33.6) $\subset$ Arg (33.7) and Arg (33.6$^{\#}$) $\subset$ Arg (33.7$^{\#}$). We consider the first inclusion. The functions that we minimize in (33.6) and (33.7) we denote as $f_1(x)$ and $f_2(x)$ respectively. These functions are concave. The sense of the number $\alpha > 0$ yields $f_1(x) \geq f_2(x)$ for all admissible x; i.e., for $x \geq 0$, $\|x\|_t \leq \beta r$. In this case, if $\tilde{x} \in$ Arg (33.6), then, taking into account the upper relation of the left-hand side of the scheme, we obtain $\|(A\tilde{x} - b)^+\|_s = \alpha$. Therefore, we have

$$\text{opt (33.6)} = f_1(\tilde{x}) = (c, \tilde{x}) - R\alpha^2 = f_2(\tilde{x}).$$

Hence it follows that $\tilde{x} \in$ Arg (33.7); therefore, the inclusion holds. The second inclusion is proved analogously. We only must take into account that the adjoint norm is piecewise linear if the norm is linear. $\square$

34. SYMMETRIC DUALITY FOR THE PARETO OPTIMIZATION PROBLEM

The Pareto optimal problem can be similarly reduced (as lexicographic problem) to the ordinary LP problem if we use scalarization. The scheme of symmetric duality is similar to those we obtain for lexicographic problem. We suppose that in problem (21.3) the vector b is represented in the form $b = Br$. Then we obtain the problem

$$P_\pi : \ \max_\pi \{C^T x \mid Ax \leq Br, \ x \geq 0\}$$

where $r > 0$ is the parameter vector. It suggested itself the following form of the dual Pareto problem

$$P_\pi^* : \ \min_\pi \{B^T u \mid A^T u \geq CR, \ u \geq 0\}.$$

Their cross scalarization gives the pair of problems mutually dual (in the classical sense):

$$P_{\pi, scal} : \ \max\{(C^T x, R) \mid Ax \leq Br, \ x \geq 0\},$$
$$P_{\pi, scal}^* : \ \min\{(B^T u, r) \mid A^T u \geq CR, \ u \geq 0\}.$$

We suppose that

$$M(r) := \{x \geq 0 \mid Ax \leq Br\}, \qquad M^*(R) := \{u \geq 0 \mid A^T u \geq CR\}.$$

Theorem 21.1 yields the following statement in this case.

Theorem 34.1. *If $M(r) \neq \varnothing$, $M^*(R) \neq \varnothing$, then the following scheme is realized:*

$$
\begin{array}{ccc}
P_\pi & \xrightarrow{\text{Arg} \atop \supset} & P_{\pi,\,scal} \\[4pt]
\updownarrow & \text{Arg} & \updownarrow \;(\#) \\[4pt]
P_\pi^* & \xrightarrow[\supset]{} & P_{\pi,\,scal}^*
\end{array}
$$

Scheme 5

In this scheme $\downarrow\uparrow$ (#) denotes the mutual correspondence in the sense of the rule (∗) of forming the dual problem in classical sense; therefore, in particular, opt $P_{\pi,\,scal} = $ opt $P_{\pi,\,scal}^*$.

Theorem 34.1, of course, remains valid if we set the problems P_π and P_π^* (analogous to (28.1) and (28.2)) in the form

$$
\max_\pi \left\{ \begin{bmatrix} c_0^T x \\ C^T x \end{bmatrix} \;\middle|\; Ax \leq b_0 + Br, \; x \geq 0 \right\},
$$

$$
\min_\pi \left\{ \begin{bmatrix} b_0^T u \\ B^T u \end{bmatrix} \;\middle|\; A^T u \geq c_0 + CR, \; u \geq 0 \right\}.
$$

Chapter 5.

Stability and well-posedness of linear programming problems

In this chapter we consider the problems of stability with respect to *value* and *argument* of the *LP* problem relatively its initial data. This problem of stability is of great importance respectively in applied mathematics. Calculation methods for solution of applied problems and organization of the process of calculation are connected with security of accuracy of the results. First of all, it is connected with accuracy of the initial data. The analysis of these difficulties is the contents of stability theory.

35. NECESSARY DEFINITIONS AND AUXILIARY RESULTS

In order to shorten the form of the *LP* problem we write it as follows:

$$L: \quad \max\{(c,x) \mid Ax \leq b\} \quad (=: \tilde{f}) \tag{35.1}$$

the dual problem will take the form

$$L^*: \quad \min\{(b,u) \mid A^T u = c, \ u \geq 0\}. \tag{35.1*}$$

The system of initial data for the problem L will be the terms which form the matrix A and the vectors b and c. The optimal value opt L of the problem L, which we denote in (35.1) as $\tilde{f}$, depends on A, b, c: $\tilde{f} = \tilde{f}(A,b,c)$. The values A, b, and c play the role of parameters when we analyze the stability of the problem L. We suppose that y is the vector consisted of the whole parameter set. For example, $y = b$, $y = c$, $y = [A, c]$. We shall consider $\tilde{f}$ as the function dependent on y: $\tilde{f} = \tilde{f}(y)$. If $y = c$ then $\tilde{f}(y) = \tilde{f}(c)$. If $y = A$, then $\tilde{f}(y) = \tilde{f}(A)$ and so on.

Definition 35.1. We shall say that solvable problem (35.1) is $\tilde{f}$-*stable* with respect to y in the point $y = y_0$ if its optimum function $\tilde{f}(y)$ is defined in a certain neighborhood of y_0 and is continuous in this point.

In this definition and further we assume that the space of variable y (suppose $\mathbb{R}^k$) is alloted by the norm

$$\|y\| = (\sum_{i=1}^{k} y_i^2)^{1/2}.$$

If $y = A$ then $\|A\| = (\sum_{j=1}^{m} \|a_j\|^2)^{1/2}$; if $y = b$ then $\|b\| = (\sum_{j=1}^{m} b_j^2)^{1/2}$ and so on.

Definition 35.2. Solvable problem (35.1) we shall call correct with respect to y in the point y_0 if $\tilde{f}(y)$ is defined in a certain neighborhood V of the point y_0 and

$$\{y_k\} \to y_0, \qquad y_k \in V, \qquad x_k \in \text{Arg}\, L(y_k)$$

yields boundedness of the sequence $\{x_k\}$ and $\{x_k\}' \subset \text{Arg}\, L(y_0)$. Here $\{x_k\}'$ is the set of limit points of the sequence $\{x_k\}$; $L(y_k)$ is the L problem for $y = y_k$.

The properties of stability and correctness of LP problems are connected essentially with stability of systems of linear inequalities.

We write the system of linear inequalities

$$Ax \le b, \qquad x \in N, \tag{35.2}$$

where N is a certain polyhedron.

Remark 35.1. Some of the parameters (of all the collection of parameters which form the matrix A and the vector b) may enter in various inequalities of system (35.2); therefore, the variations of these parameters must be coordinated. We consider, for example, the symmetric system

$$S: \quad Ax \leq b, \quad A^T u \geq c, \quad (c, x) \geq (b, u), \quad x \geq 0, \quad u \geq 0 \qquad (35.3)$$

which corresponds to the *LP* problem of form (13.5), i.e.,

$$\max\{(c, x) \mid Ax \leq b, \ x \geq 0\}. \qquad (35.4)$$

We see that all the coefficients which form the initial data of problem (35.4) are repeated. Therefore, this remark is related to the whole system of initial data.

We suppose that the sets $M := \{x \mid Ax \leq b\}$ and N in (35.2) are informatively independent. This means that in the systems of inequalities which define them we have no the connected parameters of initial data.

Definition 35.3. We suppose that y is the parameter vector of system (35.2). We shall call this system *y-stably compatible (stable)* if there exists a neighborhood V of the point y such that for each $\bar{y} \in V$ the system will be compatible. The analogous definition can be propagated on the situation when the system has equations also.

The stability of the system of linear inequalities with respect to the vector b of the right-hand sides will play a particular role. We shall call it *b*-stability (in application to system (35.3), we shall call it by *b*-stability and *c*-stability).

In order to determine the stability conditions we must establish the following moments: boundedness of the set $\operatorname{Arg} L$; existence of a neighborhood V of the point $y_0 = [A, b, c]$ such that the problem be solvable for each $y \in V$; boundedness of the set $\bigcup_{y \in V} \operatorname{Arg} L(y)$ and so on. The below statements clarify all these moments and anticipate the proof of the correspondent stability theorems.

Lemma 35.1. *For system* (35.2) *the following conditions are equivalent*

$1°$. *b-stability;*

$2°$. *$[A, b]$-stability;*

$3°$. $\exists \bar{x} \in N: \ A\bar{x} < b$.

Proof. 1° implies 3°. Really, we may take the solution $\bar{x}$ of the system $Ax \leq b + \Delta b$, $x \in N$ for $\Delta b < 0$ sufficiently small by the norm. Then, for such $\bar{x}$, condition 3° holds.

3° implies 2°. As the function vector $A\bar{x} - b$ is continuous with respect to $y = [A, b]$, then, for the points $\bar{y} = [\bar{A}, \bar{b}]$ from a certain neighborhood V of the point y the inequality $\bar{A}\bar{x} \leq \bar{b}$, $\bar{x} \in N$ will be hold. So, the condition 2° holds.

2° implies 1°. This implication is trivial. □

We consider now the problem on stability of system (35.3) in connection with connectedness of its parameters which enter into various subsystems of our system. n particular, we consider the problem on stability with respect to the whole set of its initial data, i.e., $y = [A, b, c]$.

Theorem 35.1. *System (35.3) is $[A, b, c]$-stable if and only if each of the two systems*

$$Ax < b, \quad x \geq 0 \quad \text{and} \quad A^T u > c, \quad u \geq 0 \tag{35.5}$$

is compatible.

Proof. Sufficiency. If system (35.5) is compatible, then for a certain neighborhood V of the point $y = [A, b, c]$ the systems

$$A_\delta x \leq b_\delta, \quad x \geq 0 \quad \text{and} \quad A_\delta^T u \geq c_\delta, \quad u \geq 0 \tag{35.6}$$

will be compatible for each $y_\delta := [A_\delta, b_\delta, c_\delta] \in V$. This yields solvability of the problems

$$L_\delta : \quad \max\{(c_\delta, x) \mid A_\delta x \leq b_\delta, \; x \geq 0\},$$

$$L_\delta^* : \quad \min\{(b_\delta, u) \mid A_\delta^T u \geq c_\delta, \; u \geq 0\}$$

and their optimal values coincide. Therefore, if $x_\delta \in \text{Arg}\, L_\delta$, $u_\delta \in \text{Arg}\, L_\delta^*$, then x_δ and u_δ will satisfy system (35.3) for $y = y_\delta$. This denotes that system (35.3) will be compatible for sufficiently small variations of the parameters A, b, and c. This means that the system is stable.

Necessity. If system (35.3) is $[A, b, c]$-stable, then compatibility of systems (35.5) is established analogously to the case, when we justify the implication 1° ⇒ 3° in Lemma 35.1. □

Lemma 35.2. *Suppose*

$$c = \sum_{j=1}^{k} \lambda_j a_j^T \tag{35.7}$$

where $\lambda_j > 0$, $\{a_j\}_1^k$, are linearly independent, and $\|c\| = 1$. Then $\|\lambda\| \leq C := \|(AA^T)^{-1}A\|$. Here $\lambda^T = [\lambda_1, \ldots, \lambda_k]$, $\|B\|$ is the norm of the matrix

$$B := (AA^T)^{-1}A; \; A = \begin{bmatrix} a_1 \\ \vdots \\ a_k \end{bmatrix}.$$

Proof. Equality (35.7) in the matrix form is written in the form $c = A^T \lambda$. Multiplying it by A from the left and taking into account nondegeneracy of the matrix AA^T, we obtain $\lambda = [(AA^T)^{-1}A]c$. This yields the inequality $\|\lambda\| \leq C$. □

We introduce now the operator $\pi_M(x)$ of projection of an element $x \in \mathbb{R}^n$ onto the convex closed set $M \subset \mathbb{R}^n$:

$$\pi_M(x) = \arg\min_{y \in M} \|x - y\|.$$

We can prove easily that this operator is determind uniquely and is continuous.

Further, the distance between x and M defined as $\min_{y \in M} \|x - y\|$, we shall denote by $|x - M|$.

Lemma 35.3. *Suppose that $\mathbb{R}^n \supset M \neq \varnothing$ is a convex closed set and $p \notin M$. If $\bar{p} = \pi_M(p)$, then the half-space $P := \{x \mid (p - \bar{p}, x - \bar{p}) \leq 0\}$ contains M and the hyperplane $H = \{x \mid (p - \bar{p}, x - \bar{p}) = 0\}$ is supporting to M in the point $\bar{p}$.*

Proof. The case that holds in the lemma we represent in Figure 7.

This picture clarify the formalism of the proof. If the lemma is false, then $\exists \bar{y} \in M : (p - \bar{p}, \bar{y} - \bar{p}) > 0$. If S^0 is the interior of the ball

$$S := \{x \mid \|x - p\| \leq \rho_0 := \|p - \bar{p}\|\},$$

then $S^0 \cap [\bar{p}, \bar{y}] \neq \varnothing$ (here $[\bar{p}, \bar{y}]$ is the segment which connects the points $\bar{p}$ and $\bar{y}$). We take an element $\tilde{y}$ from this intersection. Then we obtain, from

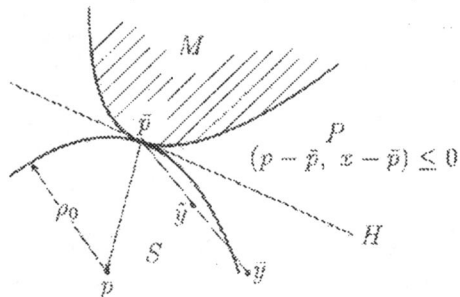

Figure 7

the one hand (since M is convex) $\tilde{y} \in M$. From the other hand, we have $\|p - \tilde{y}\| < \rho_0$ which contradicts the definition of ρ_0 (it is the distance from p to the nearest element of M). The lemma is proved. □

Lemma 35.4. *Suppose* $M = \{x \mid Ax \leq b\}$, $p \notin M$ *and* $\bar{p}$ *is the projection of* p *onto* M. *Then there exist* $\lambda_j > 0 \; j \in J \subset J(\bar{p}) := \{j \mid (a_j, \bar{p}) = b_j\}$ *such that*

$$\frac{p - \bar{p}}{\|p - \bar{p}\|} = \sum_J \lambda_j a_j^T \tag{35.8}$$

where $\{a_j\}_J$ *are linearly independent.*

Proof. We suppose that c is the left-hand side of (35.8). Then, by Lemma 35.3, the inequality $(c, x - \bar{p}) \leq 0$ is implication of the second order of the system $Ax \leq b$. By Theorem 7.1, this yields

$$(c, x - \bar{p}) \overset{(x)}{\equiv} \sum_{j \in J} \lambda_j [(a_j, x) - b_j],$$

where $\lambda_j > 0$; $\{a_j\}_J$ are linearly independent; $J \subset J(\bar{p})$. Hence the above relation follows. □

Lemma 35.5. *Suppose* $M := \{x \mid Ax \leq b\}$. *Then the inequality holds*

$$|x - M| \leq C\|(Ax - b)^+\| \tag{35.9}$$

where C *is a positive constant depending on* A *only.*

Proof. We suppose that $\bar{x}$ is a projection of $x \notin M$ onto M. Then, by Lemma 35.4, we have

$$\frac{x - \bar{x}}{\|x - \bar{x}\|} = \sum_J \lambda_j a_j^T \tag{35.10}$$

where $\lambda_j > 0$, $\{a_j^T\}_{j \in J}$ are linearly independent; $J \subset J(\bar{x})$. We suppose, for definiteness, that $J = \{1, \ldots, k\}$. Taking into account Lemma 35.2, we have $\|\lambda\| \leq C(J)$. Here $C(J)$ is the norm of the matrix $B_{(J)} = (A_J A_J^T)^{-1} A_J$;

$A_J := \begin{bmatrix} a_1 \\ \vdots \\ a_k \end{bmatrix}$. We take $c = \max_J C(J)$, which means the maximum with

respect to all collections of $\{a_j\}_J$ from Lemma 35.4 dependent on p. In our case the dependence is on x. Evidently, we have a finite number of such collections. Multiplying (35.10) by $x - \bar{x}$ scalarly, we see that

$$\|x - \bar{x}\| = \sum_J \lambda_j (a_j, x - \bar{x}) = \sum_J \lambda_j [l_j(x) - l_j(\bar{x})] = \sum_J \lambda_j l_j(x)$$

$$\leq \sum_{j=1}^m \lambda_j l_j^+(x) \leq \|\lambda\| \cdot \|(Ax - b)^+\| \leq C \|(Ax - b)^+\|.$$

The lemma is proved. $\square$

Remark 35.2. The norm $\|\cdot\|$ in (35.9) may be arbitrary. If $\|\cdot\|_p$ is another norm, then $\|\cdot\| \leq \gamma \|\cdot\|_p$. This yields the inequality

$$|x - M| \leq C_1 \|(Ax - b)^+\|_p, \qquad \text{for} \qquad C_1 = C\gamma$$

In particular, if $\|z\|_p = \max_i |z_i|$, we obtain inequality (35.9) in the form

$$|x - M| \leq C_1 \max_{(j)} l_j^+(x)$$

where $|x - M| = \inf_{y \in M} \|x - y\|$.

Lemma 35.6. *The optimal set L of the problem (35.1) is bounded if and only if the system of vectors $\{a_j^T, -c\}_1^m$ is multifold, i.e., when the system*

$$Ax \leq 0; \qquad (c, x) \geq 0; \qquad \bar{x} = 0$$

has a unique solution.

Proof. This lemma follows from Theorems 3.1 and 3.2 if we take into account the equality

$$\text{Arg}\, L = \{x \mid Ax \leq b,\ (c, x) \geq \text{opt}\, L\}.$$

$\square$

Lemma 35.7. *A multifold system of vectors $\{a_j\}_{j \in J}$ concerves the multifold property for small variations of these vectors.*

Proof. By Theorem 3.1 the property to be multifold is equivalent to the property that the system $(a_j, x) \leq 0$, $j \in J$ has a unique solution $\bar{x} = 0$. We shall prove this by contradiction. We suppose that for each $j \in J$, there exists a sequence $\{\Delta_j^k\}_k \to 0$ such that the system

$$(a_j + \Delta_j^k, x) \leq 0, \qquad j \in J \tag{35.11}$$

has a nontrivial solution $\bar{x}_k$. We may assume that $\|\bar{x}_k\| = 1$. We select from $\{\bar{x}_k\}$ a converginf subsequence: $\{\bar{x}_k\} \to s \neq 0$. Substituting $\{\bar{x}_k\}$ into (35.11) and passing to the limit with respect to k, we obtain $(a_j, s) \leq 0$, $j \in J$. As $s \neq 0$, we have obtained the contradiction. The lemma is proved. $\square$

Lemma 35.8. *Suppose the polyhedron M of the system $Ax \leq b \sim (a_j, x) \leq b_j$, $j = 1, \ldots, m$ is not empty and is bounded. If $\{A_t\} \to A$, $\{b_t\} \to b$, $t \to +\infty$, $M_t := \{x \mid A_t x \leq b_t\} \neq \varnothing$, then $\exists T$ such that the set $\bigcup_{t \geq T} M_t$ is bounded.*

Proof. By Lemma 35.7, we may take T such that for each $t \geq T$ the system $\{a_j^t\}$, $j = 1, \ldots, m$ is multifold. Therefore, the polyhedron M_t is bounded; here $A_t = \begin{bmatrix} a_1^t \\ \vdots \\ a_m^t \end{bmatrix}$. If the lemma is false, then there exists such a sequence $\{\bar{x}_t\}$, $\bar{x}_t \in M_t$ that $\{\|\bar{x}_t\|\} \to +\infty$. Dividing $A_t \bar{x}_t \leq b_t$ by $\|\bar{x}_t\|$ and passing to the limit with respect to converging subsequence $\{\bar{x}_{t_k}/\|\bar{x}_{t_k}\|\} \to s \neq 0$ we obtain $As \leq 0$. This contradicts to the property that $\{a_j\}_1^m$ is multifold. $\square$

Lemma 35.9. *Suppose the conditions of previous lemma hold. Then* $x_t \in M_t$, $\{x_t\} \to \bar{x}$ *yields* $\bar{x} \in M$.

Proof. By Lemma 35.5, we have $|x_t - M| \leq C\|(Ax_t - b)^+\|$. We shall give the following calculations:

$$(Ax_t - b)^+ = \left[(A - A_t)x_t + (b_t - b) + (A_t x_t - b_t)\right]^+ \leq \left[(A - A_t)x_t + (b_t - b)\right]^+$$

Hence it follows that

$$\|(Ax_t - b)^+\| \leq \left\|[(A - A_t)x_t + (b_t - b)]^+\right\| \to 0$$

$t \to +\infty$ Therefore $|x_t - M| \to 0$. This means that $\bar{x} \in M$. The lemma is proved. $\square$

Indeed, as the set $\bigcup_t M_t$ is bounded, than the element x_t may be taken following the rule $x_t = \arg\max_{x \in M_t} |x - M|$. By the lemma, we have $\{x_t\}' \subset M$ which yields the relation

$$\rho(M_t, M) := \max_{x \in M_t} |x - M| \to 0.$$

The value $\rho(M_t, M)$ is called semi-Hausdorf distance from the set M_t to the set M. The value

$$\rho_H(M_t, M) = \rho(M_t, M) + \rho(M, M_t)$$

is called the Hausdorf distance.

Lemma 35.10. *The optimum function* $\tilde{f}(c) = \mathrm{opt}\,(35.1)$ *with the vector* c *from problem* (35.1) *taken as the ragument, is convex in the domain of its definition. The domain of definition is convex also.*

Proof. If V is the domain of definition of $\tilde{f}(c)$, i.e. the set of those c for which (35.1) is solvable, then we verify convexity of V easily. If $c_1, c_2 \in V$ then for $\bar{x} \in M = \{x \mid Ax \leq b\}$ and $\alpha \in [0,1]$ we shall have

$$\alpha(c_1, \bar{x}) + (1 - \alpha)(c_2, \bar{x}) \leq \alpha \,\mathrm{opt}\, L|_{c=c_1} + (1 - \alpha)\,\mathrm{opt}\, L|_{c=c_2} < +\infty$$

which denotes that $\alpha c_1 + (1 - \alpha)c_2 \in V$.

We verify now convexity of $\tilde{f}(c)$:

$$\tilde{f}(\alpha c_1 + (1 - \alpha)c_2) = \max\{(\alpha c_1 + (1 - \alpha)c_2, x) \mid x \in M\}$$

$$\leq \alpha \max\{(c_1, x) \mid x \in M\} + (1 - \alpha) \max\{(c_2, x) \mid x \in M\}$$

$$= \alpha \tilde{f}(c_1) + (1 - \alpha)\tilde{f}(c_2),$$

which was to be proved.   □

Lemma 35.11. *If problem (35.1) is solvable and its optimal set* $\operatorname{Arg} L$ *is bounded, than, for small increments* Δc *of the vector* c, *the problem* $L_{\Delta c} : \max\{(c + \Delta c, x) \mid Ax \leq b\}$ *is solvable. In this case, we have*

$$\operatorname{Arg} L_{\Delta c} \subset \operatorname{Arg} L. \tag{35.12}$$

Proof.   By Lemmas 35.6 and 35.7, the system of vectors $\{a_j^T, -(c + \Delta c)\}_1^m$, will be multifold for small Δc; therefore, the polyhedron $M(t) = \{x \mid Ax \leq b, -(c + \Delta c, x) \leq -t\}$ will be bounded for any t for which $M(t) \neq \varnothing$. As the problem $L_{\Delta c}$ is equivalent to the problem

$$\max\{t \mid M(t) \neq \varnothing\} \quad (=: \bar{t})$$

where $M(t) \subset M(t')$ for $t \geq t'$, then $M(\bar{t}) \neq \varnothing$ and $\bar{t} = \operatorname{opt} L_{\Delta c}$. So we have proved that for small Δc the problem $L_{\Delta c}$ is solvable. Therefore, the optimum function $\tilde{f}(c)$ of the problem L is defined in a certain neighborhood of the point c.

We need to show inclusion (35.12). If this is not so, then $\exists \{\Delta_t c\} \to 0$: $\operatorname{Arg} L_{\Delta_t c} \not\subset \operatorname{Arg} L, \forall t$. The set $\operatorname{Arg} L_{\Delta_t c} =: \widetilde{M}_t$ is a certain k-face of the polyhedron M (Theorem 12.1). As the polyhedron M has a finite number of ??!! k-face, we may assume that $\widetilde{M}_t$ is the only $\widetilde{M}_0$ for all t.

We suppose that $\tilde{x} \in \widetilde{M} := \operatorname{Arg} L$ and $\bar{x} \in \widetilde{M}_0 := \operatorname{Arg} L_{\Delta_t c}$. The following chain of implications is evident

$$(c + \Delta_t c, \tilde{x}) \leq \tilde{f}(c + \Delta_t c) = (c + \Delta_t c, \bar{x})$$

$$\Longrightarrow (c, \tilde{x}) \leq (c, \bar{x}) \Longrightarrow (c, \tilde{x}) = (c, \bar{x}) \Longrightarrow \tilde{x} \in \operatorname{Arg} L.$$

Therefore, $\widetilde{M}_t \subset \widetilde{M}$. The lemma is proved.   □

Lemma 35.12. *If the optimal set $\widetilde{M}$ of problem (35.1) is unbounded, then there exist increments Δc of the vector c sufficiently small, so that* opt $L_{\Delta c} = +\infty$. *This means that $\tilde{f}(c)$ is not continuous in the point c.*

Proof. We suppose that $\alpha = $ opt L, then

$$\widetilde{M} = \{x \mid Ax \le b, \qquad -(c, x) \le -\alpha\},$$

As $\widetilde{M}$ is unbounded (therefore, the system $\{a_j^T, -c\}_1^m$ is not multifold), then $\exists s \neq 0 : (a_j, s) \le 0, j = 1, \ldots, m, -(c, s) \le 0$. For $\bar{x} \in M$ and Δc arbitrarily small, we provide the inequality $(c + \Delta c, s) > 0$. Then $\bar{x} + ts \in M, \forall t \ge 0$. In this case, we have $(c + \Delta c, \bar{x} + ts) = t(c + \Delta c, s) + (c + \Delta c, \bar{x}) \to +\infty$ for $t \to +\infty$. Therefore, for Δc we have chosen, opt $L_{\Delta c} = +\infty$ which was to be proved. □

36. STABILITY OF THE LINEAR PROGRAMMING PROBLEM

Stability of problem (35.1) we have defined as continuity of the function $\tilde{f}(y) = $ opt (35.1), where y is a certain part of the general informative vector $[A, b, c]$ of the initial problem. We consider the case $\tilde{f}(c)$, $\tilde{f}(b)$, and $\tilde{f}(y)$, where $y = [A, b, c]$ separately.

Theorem 36.1. *Solvable problem (35.1) is $\tilde{f}$-stable with the respect to c if and only if the optimal set* Arg L *of problem (35.1) is bounded.*

Proof. We suppose that the set Arg L is bounded. Then by Lemma 35.11, the L problem is solvable in a certain neighborhood V of the point c.

We need to show that $\tilde{f}(y)$ is continuous in the point $y = c$. There are the two ways. First, we may take into account Lemma 35.10 and Theorem 1 from Application A. The second way is to prove it directly. We choose the second way. Let $\{y_t = c + \Delta_t c\} \subset V$ and $\Delta_t c \to 0$; $x_t \in$ Arg $L_{\Delta_t c}$ and $\{x_t\} \to x'$.

We need to prove that $\tilde{f}(y_t) \to \tilde{f}(y_0)$, $y_0 = c$. By Lemma 35.11 we have $x' \in$ Arg L. Then

$$\tilde{f}(y_t) = (c + \Delta_t c, x_t) \to (c, x') = \tilde{f}(c)$$

which is sufficiency in the theorem.

Boundedness of $\operatorname{Arg} L$ is necessary for continuity of $\tilde{f}(c)$. Really, by Lemma 35.12, if $\operatorname{Arg} L$ is unbounded, there exists extremely small Δc so that the problem $L_{\Delta c}$ is unsolvable. Therefore, $\tilde{f}(c + \Delta c)$ is not defined. The theorem is proved.   □

Theorem 36.2. *Problem* (35.1) *is $\tilde{f}$-stable with respect to b is and obly if the system of restrictions $Ax \leq b$ of this problem is b-stable.*

Proof. Sufficiency. From solvability of the system $Ax \leq b + \Delta b$ for small Δb follow both solvability of the problem

$$L_{\Delta b} : \quad \max\{(c, x) \mid Ax \leq b + \Delta b\}$$

and of the dual problem also

$$L_{\Delta b}^* : \quad \min\{(b + \Delta b, u) \mid A^T u = c, \ u \geq 0\},$$

since

$$M_{\Delta b} = \{x \mid Ax \leq b + \Delta b\} \neq \varnothing, \qquad M^* = \{u \mid A^T u = c, u \geq 0\} \neq \varnothing.$$

Hence it follows that the function $\tilde{f}(b)$ is defined in a certain neighborhood V of the point b. As the function $\tilde{f}(b)$ is concave (it is verified similarly as in Lemma 35.9) it is continuous in the point (see Appendix, Theorem 1). On the other hand, b-stability of the problem L yields b-stability of the system $Ax \leq b$. The theorem is proved.   □

As for solvable L and L^* the equality $\tilde{f}(y) = \tilde{f}^*(y)$ holds for $y = c$, $y = b$ and so on, then, if we pass to the problem L^*, Theorems 36.1 and 36.2 will change their places.

Theorem 36.3. *The problem L^* is $\tilde{f}$-stable with respect to b if and only if the optimal set $ArgL^*$ of the problem L^* is bounded.*

Theorem 36.4. *The problem L^* is $\tilde{f}$-stable with respect to c is and only if the system of restrictions $A^T u = c$, $u \geq 0$ of the problem L^* is c-stable.*

We shall put in order the properties from Theorems 36.1–36.4.

$1°$. The problem L is $\tilde{f}$-stable with respect to c.

$2°$. The set $\operatorname{Arg} L$ is bounded.

$3°$. The system $A^T u = c$, $u \geq 0$ is stable.

We write the following properties for the problem L^* also

1. The problem L^* is $\tilde{f}$-stable with respect to b.

2. The set $\text{Arg}\,L^*$ is bounded.

3. The system $Ax \leq b$ is b-stable.

Summing the results of Theorems 36.1–36.4, we can formulate the following statement.

Statement. *The conditions $1°-3°$ as well as the conditions 1-3 are equivalent.*

We consider now the $\tilde{f}$-stability of the problem L with respect to $[b, c]$. The conditions of stability in this case follow from previous theorems and can be proved easily.

Theorem 36.5. *The following conditions are equivalent:*

1) L is $\tilde{f}$-stable with respect to $[b, c]$.

2) L^* is $\tilde{f}^*$-stable with respect to $[b, c]$.

3) The sets $\text{Arg}\,L$ and $\text{Arg}\,L^*$ are bounded.

4) The systems $Ax \leq b$ and $A^T u = c$, $u \geq 0$ are stable with respect to b and c respectively.

Theorem 36.6. *Solvable problem (35.1) is $\tilde{f}$-stable with respect to $[A, b, c]$ if and only if system (35.3) is stable with respect to $[A, b, c]$. Remind the system (35.3):*

$$S: \quad Ax \leq b, \quad A^T u = c, \quad u \geq 0, (c, x) \geq (b, u). \tag{36.1}$$

Proof. Note that stability of system (36.1) with respect to $[A, b, c]$ is comprehended so that the system remains compatible for coordinated variations of matrices A and A^T and b, c entering in various fragments of the system. Note also that the inequality $(c, x) \geq (b, u)$ from (36.1) we can take off because if the systems $Ax \leq b$ and $A^T u = c$, $u \geq 0$ are compatible separately, solvability of system (36.1) is guaranteed (this follows from the duality theorem for solvable LP problems). The inequality $(c, x) \geq (b, u)$ is inserted into (36.1) only in order to form the system responded to the case of symmetric problem S from Section 15 (see Remark 15.3).

Now, we begin the proof. If the problem L is $\tilde{f}$-stable with respect to $y_0 = [A, b, c]$, then the function $\tilde{f}(y)$ is defined in a certain neighborhood V of the point y_0 which yields stable solvability of system (36.1). Conversely, if system (36.1) is stable with respect to $[A, b, c]$, then L is solvable for each z from a certain neighborhood V of the point $y_0 = [A, b, c]$. We can apply Lemmas 35.8 and 35.9 to system (36.1). Namely, we shall rename the symbols of these lemmas: N is the set of solutions of the system S; N_t is the set of solutions of this system if we replace A, b, and c by A_t, b_t, and c_t with condition $y_t = [A_t, b_t, c_t] \to y_0 = [A, b, c]$. The system we denote by S_t. Stability of the system S yields solvability of the problems

$$L_t: \quad \max\{(c_t, x) \mid A_t x \leq b_t\}$$

and boundedness of the set $\bigcup_t N_t$. Problems L_t and N_t are connected as follows: $N_t = \text{Arg}\, L_t \times \text{Arg}\, L_t^*$ (see Theorem 15.5). If

$$\{[x_t, u_t]\} \to [\bar{x}, \bar{u}], \qquad x_t \in \text{Arg}\, L_t, \qquad u_t \in \text{Arg}\, L_t^*,$$

then, by Lemma 35.9, $[\bar{x}, \bar{u}] \in N$ which means that $[\bar{x}, \bar{u}]$ is a solution of the system S. Therefore, $\bar{x} \in \text{Arg}\, L$, $\bar{u} \in \text{Arg}\, L^*$. Thus, we have

$$\tilde{f}(y_t) = (c_t, x_t) \to (c, \bar{x}) = \tilde{f}(y_0).$$

This denotes that the function $\tilde{f}(y)$ is continuous in the point y_0. □

As each of the conditions 1)-4) of Theorem 36.5 provides stability of the system S with respect to $[A, b, c]$, the following theorem holds.

Theorem 36.7. *The problem L is $\tilde{f}$-stable with respect to the whole system of initial data $[A, b, c]$ if and only if at least one of the conditions 1)-4) of Theorem 36.5 holds.*

Remark 36.1. System (36.1) is the symmetric system S of type (35.3) correlated with the LP problem in form (35.1). The stability condition for system (35.3) was formulated in Theorem 35.1. We do not consider the stability conditions for system (36.1). However, such condition may be expressed, for example, in the form:

System (36.1) *is* $[A, b, c]$-*stable is and only if the system* $Ax < b$ *is compatible and the system* $Ax \leq 0$, $(c, x) \geq 0$ *has a unique (trivial) solution.*

Really, uniqueness of solution of the system $Ax \leq 0$, $(c, x) \geq 0$ is equivalent to the multifold property of the system $\{a_j^T, j = 1, \ldots, m; -c\}$ which defined the boundedness of the set $\operatorname{Arg} L$. By Lemmas 35.1 and 35.6, small variations A, b, and c do not violate solvability of problems L and L^*. This provides the property of solvability of system (36.1) for these variations. Necessity of these conditions can be established easily also.

37. WELL-POSEDNESS OF LINEAR PROGRAMMING PROBLEMS

Notions of stability and correctness introduced above do not coincide in general case for the problems of mathematical programming. We shall show below that for the *LP* problems these notions coincide. Of course, the notions of stability (with respect to the value) and correctness (with respect to the argument) we may weaken. In this case they will not be equivalent. In our Definitions 35.1 and 35.2, the essential requirement is the existence of such neighborhood V of the point y_0 that $L(y)$ is solvable for each $y \in V$. We can weaken these definitions as follows: we can take certain fixed convex set $V \ni y_0$ and require solvability of $L(y)$ for $y \in V$. In particular, we can set (for $y = b$):
$$V = \{b + \Delta b \mid \Delta b \geq 0, \ \|\Delta b\| \leq 1\}.$$

Then, even in this particular case stability will not be equivalent to correctness.

Below we shall assume that stability and correctness is comprehended in the sense of Definitions 35.1 and 35.2.

We introduce the notions $y_t = [A_t, b_t, c_t]$ and $y_0 = [A, b, c]$. In system (36.1) we replace A, b, c by A_t, b_t, c_t, $t = 1, 2, \ldots$:

$$S_t : \ A_t x \leq b_t, \quad A_t^T u = c_t, \quad u \geq 0, \quad (c_t, x) \geq (b_t, x). \tag{37.1$_t$}$$

Note, that the set of solutions of system (37.1$_t$) coincides with $\operatorname{Arg} L_t \times \operatorname{Arg} L_t^*$, where L_t is the problem L with $[A, b, c]$ changed by $[A_t, b_t, c_t]$.

We consider the relations

$$\{y_t\} \to y_0, \qquad \{x_t \in \operatorname{Arg} L_t\} \to x' \tag{37.2}$$

which will be used below when proving the theorem on correctness of problem (35.1).

Theorem 37.1. *Solvable problem (35.1) is correct with respect to the whole system of initial data $y_0 = [A, b, c]$ if and only if it is $\tilde{f}$-stable with respect to y in the point y_0. The theorem holds also $y_0 = b$ and $y_0 = c$.*

Proof. We shall give the scheme of the proof. As it follows from the definitions, correctness yields $\tilde{f}$-stability. Conversely, if the problem is $\tilde{f}$-stable, then in order to prove that the problem is correct we need to show (in the case of relations (37.2)) that $x' \in \operatorname{Arg} L$. We have $\tilde{f}(y_t) \to \tilde{f}(y_0)$; therefore,

$$\tilde{f}(y_t) = (c_t, x_t) \to \operatorname{opt} L(= \tilde{f}(y_0)).$$

On the other hand, $(c_t, x_t) \to (c, x')$, i.e. $(c, x') = \operatorname{opt} L$. As $x' \in M = \{x \mid Ax \le b\}$, then $x' \in \operatorname{Arg} L$ which was to be proved. $\square$

38. THE TIKHONOV REGULARIZATION OF LINEAR PROGRAMMING PROBLEMS

Suppose, we are given unstable LP problem L:

$$L : \max\{(c, x) \mid Ax \le b\} \quad (=: \tilde{f}) \tag{38.1}$$

then the goal of regularization is transformation of this problem to the other problem, suppose P, which would be stable. In this case, we must have $\operatorname{Arg} P \subset \operatorname{Arg} L$.

In such construction we have strengthen the requirements to the problem P. Usually the problem P is constructed as $P = P_\alpha$, where $\alpha > 0$ is a certain parameter. The less is α, the more precisely the problem P_α approximates the problem L.

Regularization of problem (38.1) by Tikhonov is the change of this problem by the problem

$$P_\alpha : \max\{(c, x) - \alpha \|x - p\|^2 \mid x \in M\} \quad (=: \tilde{f}_\alpha), \tag{38.2}$$

where $M = \{x \mid Ax \le b\}$; p is a certain fixed point from $\mathbb{R}^n$ and $\alpha > 0$. We assume that $p \notin M$.

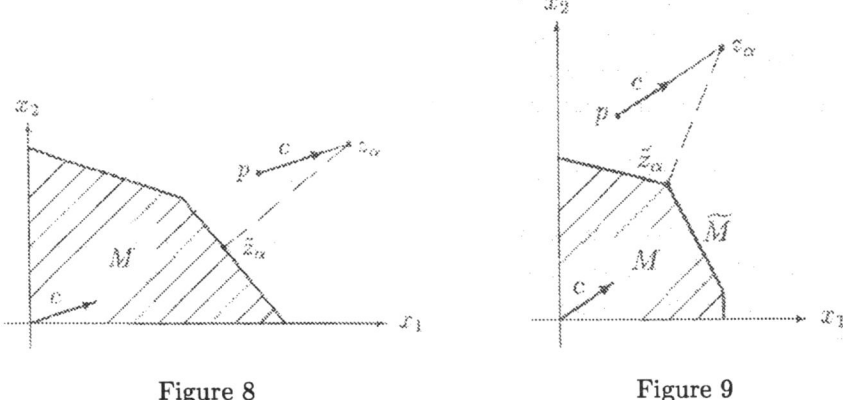

Figure 8   Figure 9

38.1. Geometric sense of the Tikhonov regularization

The following equality can be verified immediately:

$$(c, x) - \alpha \|x - p\|^2 = -\alpha \left\| x - \left(p + \frac{c}{2\alpha} \right) \right\|^2 + (c, p) + \frac{1}{4\alpha} \|c\|^2.$$

This equality yields

$$\tilde{f}_\alpha = -\alpha \operatorname{opt} \min \left\{ \left\| x - \left(p + \frac{c}{2\alpha} \right) \right\|^2 \ \middle| \ x \in M \right\} + (c, p) + \frac{1}{4\alpha} \|c\|^2.$$

In this case

$$\arg P_\alpha = \arg \min_{x \in M} \left\| x - \left(p + \frac{c}{2\alpha} \right) \right\|^2 \quad (= \tilde{z}_\alpha). \tag{38.3}$$

We set $z_\alpha = p + \frac{c}{2\alpha}$. The right-hand side in (38.3) is the projection $\tilde{z}_\alpha$ of the point z_α onto M, which is the nearest to z_α element from M. Thus, the optimal vector $\tilde{z}_\alpha$ of the problem P_α is the projection of the vector $z_\alpha = p + \frac{c}{2\alpha}$ onto the admissible set M of the problem L (see Figure 8).

Indeed, for $\alpha > 0$ sufficiently small, $\tilde{z}_\alpha$ will coincide with the projection of the point p onto $\operatorname{Arg} L =: \widetilde{M}$ (see Figure 9), i.e.

$$\arg P_\alpha = \arg \min_{p \in \widetilde{M}} \|x - p\|. \tag{38.4}$$

We need to prove relation (38.4).

38.2. An auxiliary lemma

The following problem is more general comparatively the problem P_α:

$$F_\alpha : \max\{(c, x) - \alpha f(x) \mid Ax \le b\}. \tag{38.5}$$

Here $f(x)$ is a convex function where the Lebesque sets $\{x \mid f(x) \le \beta\}$ are bounded. The function $\|x - p\|^2$ is an example of such function. Parallel with (38.5) we consider the two-step optimization problem

$$\min\{f(x) \mid x \in \operatorname{Arg} L\}. \tag{38.6}$$

Lemma 38.1. *For $\alpha > 0$ sufficiently small, we have*

$$\operatorname{Arg} F_\alpha = \operatorname{Arg}(38.6). \tag{38.7}$$

Proof. In the sace of problem (38.6) we can use the method of exact penalty functions. Namely, we rewrite (38.6) in the form

$$\min\{f(x) \mid Ax \le b, (c, x) \ge \tilde{f}\}, \tag{38.8}$$

where $\tilde{f} = \operatorname{opt} L$. Using the scheme of exact penalty functions (see Section 20 and also Appendix, Theorems 8 and 15), problem (38.8), i.e. problem (38.6) is equivalent (we mean coincidence of optimal values and optimal sets) to the problem

$$\min\{f(x) + R_0[\tilde{f} - (c, x)]^+ \mid x \in M\}, \tag{38.9}$$

where $R_0 > |\bar{u}_0|$. Here $\bar{u}_0$ is the dual estimate of the inequality $(c, x) \ge \tilde{f}$ in problem (38.8). As for $x \in M$, we have $\tilde{f} \ge (c, x)$; then we can remove the cut "+" in (38.9). Therefore, we can replace (38.9) by

$$\min\{f(x) + R_0(c, x) + R\tilde{f} \mid x \in M\}. \tag{38.10}$$

Hence it follows that

$$\operatorname{Arg}(38.6) = \operatorname{Arg}(38.8) = \operatorname{Arg}(38.10)$$

$$= \operatorname{Arg} \max \left\{(c, x) - \frac{1}{R_0} f(x) \;\middle|\; x \in M\right\},$$

which gives (38.7) for $\alpha = R_0^{-1}$. The lemma is proved. □

If $f(x) = \|x - p\|^2$, then the lemma denotes relation (38.4). As we are interested in this relation, we formulate the following statement.

Theorem 38.1. *If problem (38.1) is solvable, then, for $\alpha > 0$ sufficiently small, the unique solution of problem (38.2) (i.e., of the problem P_α) coincides with projection of the vector p onto the optimal set $\mathrm{Arg}\,L$ of the problem L.*

38.3. Tikhonov regularization of improper linear programming problems

We consider first *IPLP* problem of the second order. Remind that *LP* problem, suppose (38.1), of the second order is the case when $M = \{x \mid Ax \leq b\} \neq \varnothing$, but $\sup_{x \in M}(c, x) = +\infty$. Nevertheless, in this case, the problem P_α is solvable and its solution $\tilde{z}_\alpha$ is interpreted as projection of the vector z_α onto M (see Figure 10).

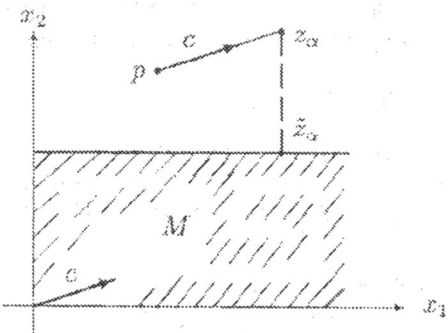

Figure 10

We suppose that the problem L is an improper problem of the first order, i.e., $M = \varnothing$, $M^* := \{u \geq 0 \mid A^T u = c\} \neq \varnothing$. The following two problems which realize the particular situation of $(\#)$-duality (see Section 26):

$$P: \quad \max_{(x)}\{(c, x) - (R, (Ax - b)^+)\}, \tag{38.11}$$

$$P^\#: \quad \min\{(b, u) \mid A^T u = c, 0 \leq u \leq R\} \tag{38.12}$$

are corresponded to the L problem and the dual problem $L^* : \min_{u \in M^*}(b, u)$ respectively.

By Theorem 25.1, these problems are solvable for each $R > 0$ and their optimal values coincide. If we regularize these problems by Tikhonov, we obtain the problems

$$P_\alpha: \quad \max_{(x)}\{(c, x) - \alpha\,\|x - p\|^2 - (R, (Ax - b)^+)\},$$

$$P_\beta^\# : \min_{(u)}\{(b, u) + \beta\|u - q\|^2 \mid A^T u = c, \ 0 \leq u \leq R\},$$

where p and q are fixed points from $\mathbb{R}^n$ and $\mathbb{R}^m$ respectively, $\alpha > 0$, $\beta > 0$. Interpretation of these problems may be various depending on the value of parameter $R > 0$. For example, we suppose that $R = \gamma\bar{R}$, $\bar{R} > 0$. We consider the three-step problem of subsequence programming

$$\min(\bar{R}, (Ax - b)^+), \tag{38.13$_1$}$$

$$\max\{(c, x) \mid x \in \text{Arg}\,(38.13_1)\}, \tag{38.13$_2$}$$

$$\boxed{\min\{\|x - p\|^2 \mid x \in \text{Arg}\,(38.13_2)\}.} \tag{38.13}$$

Following the method of exact penalty functions, we can choose $\gamma > 0$ such that $\text{Arg}\,P = \text{Arg}\,(38.13_2)$. For $\alpha > 0$ sufficiently small, we shall have $\arg P_\alpha = \arg\,(38.13)$, which means that the optimal vector of the regularized problem P_α will be the optimal vector of the problem of subsequent programming (38.13).

The problem $P_\beta^\#$ is interpreted simpler: for $\beta > 0$ sufficiently small, the optimal vector of the problem $P_\beta^\#$ coincides with the projection of the point q onto the set $\text{Arg}\min\{(b, u) \mid A^T u = c, \ 0 \leq u \leq R\}$.

So, the problem P_α is regularizing with respect to P, i.e. to (38.11). This problem is equivalent to problem (38.13) (by the optimal vector). Moreover, the reduction $P \to P_\alpha$ satisfies the conditions of stability of P_α with respect to the initial data. Of course, we must justify some moments here. However, if, for example, we consider stability with respect to the vector c, it holds since the projection operator

$$\pi_M(x) : x \to \arg\min_{y \in M} \|x - y\|$$

is continuous. Really, we have noted yet that the problem of maximization of the function $(c, x) - \alpha\|x\|^2$ in M is equivalent to projection of the vector $p + \frac{c}{2\alpha}$ on M. In our case, $M = \text{Arg}\min_{(x)}(R, (Ax - b)^+)$. All this is similar for the problem $P_\beta^\#$.

For completeness, we consider the symmetric regularized problems for L and L^*. We shall not define more exactly the order of their improperness and take them in forms (15.1) and (15.1*):

$$L : \quad \max\{(c, x) \mid Ax \leq b, \ x \geq 0\},$$

$$L^* : \quad \min\{(b, u) \mid A^T u \geq c, \ u \geq 0\}.$$

If we make ($\#$)-reduction of improper problems L and L^* to the proper problems C and $C^\#$ respectively (26.1) and (26.1$^\#$), then the analogues of P and P_α here will be the problems

$$Q_\alpha : \quad \max\{(c, x) - \alpha \, \|x - p\| - (R, (Ax - b)^+) \mid 0 \leq x \leq r\},$$

$$Q_\beta^\# : \quad \min\{(b, u) + \beta \, \|u - q\| + (r, (c - A^T u)^+) \mid 0 \leq u \leq R\}.$$

As problems (26.1) and (26.1$^\#$) are solvable without aby assumptions, and their optimal values coincide, then the problems Q_α and $Q_\beta^\#$ are solvable also. In this case, we have

$$0 \leq \operatorname{opt} Q_\beta^\# - \operatorname{opt} P_\alpha^\# \leq \alpha \, \|\bar{x} - p\| + \beta \, \|\bar{u} - q\|,$$

where $\bar{x}$ is the projection p onto $\operatorname{Arg}(26.1)$; $\bar{u}$ is the projection q onto $\operatorname{Arg}(26.1^\#)$; $\alpha > 0$ and $\beta > 0$ are sufficiently small. If we consider problem (26.1) as the problem of optimal correction of the problem L (i.e., the problem L is given approximately, $y_\delta = [A_\delta, b_\delta, c_\delta]$, $\|y_\delta - y\| \leq \delta$, $y = [A, b, c]$), then $\|\bar{x}_\delta - \bar{x}\| \to 0$, and $\|\bar{u}_\delta - \bar{u}\| \to 0$ for $\delta \to 0$. Here $\bar{x}_\delta$ and $\bar{u}_\delta$ are the solutions of the regularized problems $Q_{\alpha,\delta}$ and $Q_{\beta,\delta}^\#$ respectively. The problems $Q_{\alpha,\delta}$ and $Q_{\beta,\delta}^\#$ are the problems Q_α and $Q_\beta^\#$ where A, b, and c are replaced by A_δ, b_δ, and c_δ.

38.4. Stable unsolvability of LP problems

Unsolvability (incompatibility) of systems of linear inequalities and unsolvability (improperness) of LP problems are important as well as solvable problems. They are characterized by their own problems and contents. We consider here the problems of their *stable unsolvability*. We mean here the conservation of unsolvability for small variations of the initial data. We write the system of linear inequalities

$$Ax \leq b. \tag{38.14}$$

We shall call this system *stably unsolvable* if it is unsolvable for small variations of the matrix A and the vector b. As unsolvable LP problems have three orders, we can introduce the notion of stability which conserves the order of unsolvability for small variations of initial data.

As for stability of inequality systems and *LP* problems in the case of their solvability, these questions where considered in previous sections. In particular, stability of system (38.14), if it is compatible, is equivalent to solvability of system of rigorous linear inequalities $Ax < b$. For system $Ax \leq b$, $x \geq 0$, the analogous condition will be solvability of the system $Ax < b$, $x \geq 0$.

Theorem 38.2. *Incompatible system* (38.14) *is stably incompatible is and only if the system* $Ax \leq 0$ *has a unique solution (trivial solution).*

Proof. Recall, (Theorem 3.1) that condition of uniqueness of solution of the system $Ax \leq 0$ is equivalent of multifold property for the system of row vectors of the matrix A (or, to the property that this system of rows is a non-negative basis in $\mathbb{R}^n$).

So, we suppose that (38.14) is incompatible and the system $Ax \leq 0$ has a unique (trivial) solution. We show that system (38.14) is stable incompatible. If this is not so, we can choose such $\{A_k\}$, $\{b_k\}$ that $\{A_k\} \to A$, $\{b_k\} \to b$ and each system $A_k x \leq b_k$ will have the solution x_k. We may assume that $\{x_k/\|x_k\|\} \to s$, $\|s\| = 1$. If the sequence $\{x_k\}$ is unbounded then from $A_k(x_k/\|x_k\|) \leq b_k/\|x_k\|$ follows $As \leq 0$. This contradicts the condition. If $\{x_k\}$ is bounded (we may consider $x_k \to \bar{x}$), then $A_k x_k \leq b_k$, $\forall k$ yields $A\bar{x} \leq b$ which contradicts unsolvability of system (38.14).

We suppose now that system (38.14) is stably incompatible. We shall prove now that the system of rows vectors $\{a_j\}_1^m$ of the matrix A is a non-negative basis of the space $\mathbb{R}^n$. By Theorem 8.1, stable incompatibility of system (38.14) denotes stable compatibility of the system

$$A^T u = 0, \qquad u \geq 0, \qquad (b, u) < 0. \tag{38.15}$$

If $u^T = [u_1, \ldots, u_m]$, then in (38.15) we can take $\sum_{j=1}^m u_j = 1$. The equation $A^T u = 0$ we write in the form $\sum_{j=1}^m a_j^T u_j = 0$. As system (38.15) is stable for small $x \in \mathbb{R}^n$, the system

$$\sum_{j=1}^m (a_j^T - x) u_j = 0, \qquad \sum_{j=1}^m u_j = 1, \qquad u \geq 0$$

will be solvable respectively u, i.e. $\sum_{j=1}^m (a_j^T - x)\bar{u}_j = 0$, $\bar{u} \geq 0$, where $\bar{u}$ is its solution. Hence it follows that $\sum_{j=1}^m a_j^T \bar{u}_j = x$; therefore, $x \in \text{cone}\{a_j^T\}$. In this inclusion the proximity of the norm of x is indifferent, this inclusion holds for each $x \in \mathbb{R}^n$. This denotes that $\text{cone}\{a_j^T\} = \mathbb{R}^n$, i.e.,$\{a_j^T\}$ is a nonnegative basis of the space $\mathbb{R}^n$ (as it was noted, this is equivalent to unique solvability of the system $Ax \leq 0$). The theorem is proved. □

Conditions of stable solvability and unsolvability of systems of linear inequalities allows us to formulate conditions of stable unsolvability for *LP* problems. In particular, we can take two account the orders of their unsolvability (improperness).

Theorem 38.3. *If L is $IPLP$ problem of the first order then it is stable relatively this property if and only if the system $A^T u \geq c$, $u \geq 0$ is stably solvable (which is equivalent to solvability of the system $A^T u > c$, $u \geq 0$). The system $Ax \leq b$, $x \geq 0$ is stably unsolvable (which is equivalent to unsolvability and unique solvability of system $Ax \leq 0$, $x \geq 0$).*

The theorem follows from the conditions of stable solvability and stable unsolvability for systems of linear inequalities (Theorem 38.2) and classifying properties of improper problems of linear programming. This holds also for formulations of the Theorems 38.4–38.6 below.

Theorem 38.4. *If a problem L is an $IPLP$ problem of the second order, then it is stable relatively this property if and only if the system $Ax \leq b$, $x \geq 0$ is stably compatible (it is equivalent to compatibility of the system $Ax < b$, $x \geq 0$). The system $A^T u \geq c$, $u \geq 0$ is stably incompatible (this is equivalent to incompatibility and unique stability of system $A^T u \geq 0$, $u \geq 0$).*

Theorem 38.5. *If a problem L is $IPLP$ problem of the third order, it is stable relatively this property if $u \geq 0$ are stably incompatible (this is equivalent to their incompatibility and uniqueness of solution of each system $Ax \leq b$, $x \geq 0$ and $A^T u \geq 0$, $u \geq 0$).*

Theorem 38.6. *The problem L is stably unsolvable if and only if at least one of the systems $Ax \leq b$, $x \geq 0$ and $A^T u \geq c$, $u \geq 0$ is stably unsolvable.*

38.5. Conjunction of the Tikhonov regularization methods and square penalty functions

The method of square penalty functions for the problem

$$L: \ \max\{(c, x) \mid Ax \leq b, \ x \geq 0\}$$

is the reduction of this problem to the following one

$$P_\alpha : \max_{x \geq 0} \Phi(\alpha, x),$$

where $\Phi(\alpha, x)$ may be, for example, the function of form (20.5); namely:

$$\Phi(\alpha, x) = (c, x) - \sum_{j=1}^{m_0} \frac{1}{4\alpha_j} \|(A_j x - b^j)^+\|^2, \qquad \alpha_j > 0, \qquad j = 1, \ldots, m_0.$$

All the norms here and further are Euclidian. We shall follow the scheme

$$
\begin{array}{ccc}
L & \longrightarrow & P_\alpha \\
\updownarrow & & \updownarrow \ (*) \\
L^* & \xrightarrow{\ \text{reg}\ } & (L^*)_{\text{reg}} : \min\Big\{ (b, u) + \sum_{j=1}^{m_0} \alpha_j \|u^j\|^2 \ \Big| \ A^T u \geq c, \ u \geq 0 \Big\}.
\end{array}
$$

Scheme 6

The symbol "reg" here denotes the Tikhonov regularization; $\{u_j\}_1^{m_0}$ are the fragments of the vector u correspondent to partition of the vector (of linear functions) $Ax - b$ into the subvectors $A_j x - b^j$, $j = 1, \ldots, m_0$.

We shall establish here that although the problem P_α solves the problem L (see Theorem 20.2) approximately, however, the problem $(L^*)_{\text{reg}}$ dual to P_α solves L^* exactly for α_j sufficiently small. *

We consider first the problem of constructing the dual problem to those which regularize the problem L:

$$\max\Big\{ (c, x) - \sum_{i=1}^{n} \beta_i x_i^2 \ \Big| \ Ax \leq b, \ x \geq 0 \Big\}. \tag{38.16}$$

Following general rule of duality forming (see Appendix, Section A3), the problem dual to (38.16) will be as follows:

$$\min_{u \geq 0, v \geq 0} \max_{(x)} F(x, u, v), \tag{38.17}$$

---

*Although this fact follows immediately from the above results on the penalty functions obtained by the (Eremin, 1988), however, the author has no exact formulation of this fact. The formulation was suggested by prof. A.I. Golikov in 1995.

where

$$F(x, u, v) = (c, x) - \sum_{i=1}^{n} \beta_i x_i^2 - (u, Ax - b) + (v, x)$$

is the Lagrange function correspondent to problem (38.16). We can change this problem by the equivalent problem

$$\min\{F(x, u, v) \mid \nabla_x F(x, u, v) = 0, \ u \geq 0, \ v \geq 0\} \tag{38.18}$$

(see (7) and (8*), Appendix, Section A3). Substituting $F(x, u, v)$ into eqref 38.18, we obtain

$$\min \Big\{ (c - A^T u + v, x) - \sum_{i=1}^{n} \beta_i x_i^2 + (b, u) \ \Big|$$
$$c - A^T u + v = 2 \sum_{i=1}^{n} \beta_i x_i, \ u \geq 0, v \geq 0 \Big\}.$$

Finally, we obtain

$$\min \Big\{ (b, u) + \sum_{i=1}^{n} \frac{1}{4\beta_i} [c_i - (h_i, u) - v_i]^2 \ \Big| \ u \geq 0, \ v \geq 0 \Big\}, \tag{38.19}$$

where $\{c_i\}_1^n$ are coordinates of the vector c; $\{v_i\}_1^n$ are the coordinates of the vector v; $\{h_i\}_1^n$ are the column vectors of the matrix A. We can remove the vector $v \geq 0$ from (38.19) if we replace $[c_i - (h_i, u) - v_i]^2$ by $[c_i - (h_i, u)]^{+2}$. This can be done since $\min_{v_i \geq 0} |z - v_i| = z^+$. Then, for the problem (38.18) dual to (38.16), we obtain the final form

$$\min \Big\{ (b, u) + \sum_{i=1}^{n} \frac{1}{4\beta_i} [c_i - (h_i, u)]^{+2} \ \Big| \ u \geq 0 \Big\}. \tag{38.20}$$

Thus, we have reduced the problem L^* to problem (38.20) following the method of squared penalty functions. So, we have connected the problems (38.16) and (38.20) by duality. Taking into account Theorem 12 from Appendix, we state that the optimal values of these problem coincide, i.e.,

$$\text{opt}\,(38.16) = \text{opt}\,(38.20). \tag{38.21}$$

Instead of problem (38.16) we may consider the problem

$$\max\left\{(c,x) - \sum_{i=1}^{n_0}\beta_i\|x^i\|^2 \;\middle|\; Ax \le b,\; x \ge 0\right\}, \qquad (38.22)$$

as a particular, where $\{x^i\}_1^{n_0}$ is an arbitrary partition of the vector x into the subvectors x^i, $i = 1, \ldots, n_0$. Then, the dual problem (i.e. the analog of problem (38.20)) will be as follows:

$$\min\left\{(b,u) + \sum_{i=1}^{n_0}\frac{1}{4\beta_i}\|(c^i - B_i^T u)^+\|^2 \;\middle|\; u \ge 0\right\}, \qquad (38.23)$$

where $\{c^i\}_1^{n_0}$ is the partition to the partition of x into the fragments c_i correspondent to the partition of x into $\{x^i\}_1^{n_0}$; $\{B_i\}_1^{n_0}$ is correspondent partition of the matrix A into vertical submatrices.

If we apply this scheme of transfer to dual problem for the problem $(L^*)_{\text{reg}}$ from the Scheme 6, we obtain just the problem P_α – the analog of problem (38.23). Therefore, we have coincidence of optimal values of the problem $(L^*)_{\text{reg}}$ and P_α.

We shall be interested in problems L, L^*, $(L^*)_{\text{reg}}$, and P_α in totality. Their mutual properties are as follows:

1°. $\operatorname{opt} L = \operatorname{opt} L^* =: \tilde{m}$ (dual theorem in linear programming).

2°. Suppose $\tilde{u} \in \operatorname{Arg} L^*$, $\{\tilde{u}_j\}_1^{m_0}$ are fragments of the vector $\tilde{u}$ correspondent to the partition of A into A_j, $j = 1, \ldots, m_0$; $m_\alpha := \operatorname{opt} P_\alpha$. Then $|m_\alpha - \tilde{m}| \le \sum_{j=1}^{m_0}\alpha_j\|\tilde{u}^j\|^2$. (This follows from estimate (20.6)).

3°. $\operatorname{opt} P_\alpha = \operatorname{opt}(L^*)_{\text{reg}}$ (this is analog of relation (38.21)).

4°. We suppose that v_0 is the dual estimate of the inequality $(b,u) \le \tilde{m}$ in the problem

$$\min\{\|u\|^2 \;|\; A^T u \ge c,\; u \ge 0,\; (b,u) \le \tilde{m}\}.$$

Then, for $\alpha_j = \bar{\alpha}_j t$ and $t^{-1} > |v_0|\bar{\alpha}_j$, $j = 1, \ldots, m_0$, the relation

$$\arg(L^*)_{\text{reg}} = \tilde{u} \in \operatorname{Arg} L^*$$

holds, where

$$\tilde{u} := \arg\min\Big\{ \sum_{j=1}^{m_0} \bar{\alpha}_j \|\tilde{u}^j\|^2 \ \Big|\ u \in \operatorname{Arg} L^* \Big\} \tag{38.24}$$

(particular case Lemma 38.1 applied to this situation).

From this properties this statement below follows.

Theorem 38.7. *Suppose the problem L is solvable and reduced to P_α by the method of square penalty functions with the penalty constants $\alpha_j = \bar{\alpha}_j t$, $\bar{\alpha}_j > 0$, $j = 1, \ldots, m_0$. Then, for $t^{-1} > |v_0|\bar{\alpha}_j$, $j = 1, \ldots, m_0$, the solution of the problem $(L^*)_{\mathrm{reg}}$ is unique and coincides with $\tilde{u}$ from (38.24). In this case*

$$|m_\alpha - \tilde{m}| = \sum_{j=1}^{m_0} \alpha_j \|\tilde{u}^j\|^2. \tag{38.25}$$

Proof. Really, by the property $3°$, we have $m_\alpha = (b, \bar{u}) + \sum_{j=1}^{m_0} \alpha_j \|\tilde{u}^j\|^2$. Moreover, the property $4°$ yields

$$\tilde{u} := \arg\min\Big\{ \sum_{j=1}^{m_0} \alpha_j \|\tilde{u}^j\|^2 \ \Big|\ u \in \operatorname{Arg} L^* \Big\}.$$

Therefore, taking into account $(b, \tilde{u}) = \tilde{m}$, we obtain the desired relation (38.25). The theorem is proved.   □

Remark 38.1. We see that, from the one hand, the estimate $|m_\alpha - \tilde{m}| \le \sum_{j=1}^{m_0} \alpha_j \|\tilde{u}^j\|^2$ holds for each $\alpha_j > 0$, $j = 1, \ldots, m_0$. From the other hand, if $\min_{(j)} \alpha_j^{-1} > |v_0|$, then the estimate of inclination of opt P_α from opt L attains the exact form of equality (38.25).

Remark 38.2. If we take the function $\Phi(\alpha, x)$ as follows:

$$\Phi_1(\alpha, x) = (c, x) - (4\alpha)^{-1}\|(Ax - b)^+\|^2$$

which corresponds to $m_0 = 1$, then the problem $(L^*)_{\mathrm{reg}}$ will take the form

$$\min\{(b, u) + \alpha \|u\|^2 \mid A^T u \geq c, \ u \geq 0\}. \tag{38.26}$$

In this case the problem

$$P_\alpha^1 : \ \max \Phi_1(x, \alpha)$$

is dual for (38.26). Theorem 38.7 for the selected form (more simple formulation) will be as follows:

If the problem L is solvable and $\alpha^{-1} > |v_0|$, then $|\mathrm{opt}\, P_\alpha^1 - \mathrm{opt}\, L| = \alpha \|\tilde{u}\|^2$, where $\tilde{u}$ is the nornal solution of the problem L^*.

The below remark will be connected with a certain modification of the stabilizing function for regularization of *LP* problem. Instead of $\sum_{i=1}^n \beta_i \|x^i\|^2$ (or $\beta\|x\|^2$ in the simplest case) in (38.22) we take. $\beta x^T Q x$, where Q is positive definite matrix, $\beta > 0$. In this situation problems (38.22) and (38.23) will be as follows:

$$\max\{(c, x) - \beta x^T Q x \mid Ax \leq b, \ x \geq 0\}, \tag{38.27}$$

$$\min_{u \geq 0} \left\{ (b, u) + \frac{1}{4\beta}(c - A^T u)^{+T} Q^{-1}(c - A^T u)^+ \right\}. \tag{38.28}$$

In this case, problem (38.28) is dual to (38.27). This yields the equality opt (38.27) = opt (38.28) (if the problem L is solvable).

If we take the problem

$$\max_{x \geq 0} \left\{ (c, x) - \frac{1}{4\beta}(Ax - b)^{+T} Q(Ax - b)^+ \right\}, \tag{38.29}$$

as initial, then the dual pre-image to this problem will be the problem

$$\min\{(b, u) + \beta u^T Q^{-1} u \mid A^T u \geq c, \ u \geq 0\}. \tag{38.30}$$

So, problem (38.30) will be dual to problem (38.29), which defines coincidence of their optimal values (of course, if the problem L is solvable).

For problems (38.29) and (38.30) we have the following analog of Theorem 38.7.

Theorem 38.8. *Suppose the problem L is solvable. Then, for $\beta^{-1} > |v_0|$ the solution of problem (38.30) is unique and coincides with solution of the problem*

$$\min\{u^T Q^{-1} u \mid u \in \text{Arg } L^*\} =: \tilde{u}. \qquad (38.31)$$

In this case, $|m_\beta - \tilde{m}| = \beta \tilde{u}^T Q^{-1} \tilde{u}$. Here v_0 is dual estimate of the inequality $(b, u) \leq \tilde{m}$ in problem (38.30), where the set $\text{Arg } L^*$ is written in the form of the system of linear inequalities $A^T u \geq c$, $u \geq 0$, $(b, u) \leq \tilde{m}$ $m_\beta :=$ opt (38.29).

Chapter 6.

Methods of projection in linear programming

In Chapter 5 we have used the operation of projection of an element $x \in \mathbb{R}^n$ on a convex closed set $M \subset \mathbb{R}^n : x \xrightarrow{\pi_M(x)} \arg\min_{y \in M} \|x - y\|$. This operation is often used in mathematics, particularly, in applied mathematics. This operation may be considered in more general metric spaces and it is the pivot of certain mathematical directions, for example, of theory of functional approximations. For some convex sets from $\mathbb{R}^n$ this operation is rather simple. These convex sets are: $\mathbb{R}^n_+$-non-negative orthant, hyperplane, linear manifold, half-space, ball, parallelepiped and so on. Using these simplest types of projection operators we can build more complicated projections which lie in the basis of construction of the methods for solving more complicated problems. For example, if $\pi_j(x)$ is the projection on j-th half-space P_j of the system of linear inequalities $Ax \leq b$, then taking there superposition, we obtain the mapping $\varphi(x) = \pi_1 \cdots \pi_m(x)$. This mapping is such that

$$\{\varphi^t(x_0)\}_{t=1}^{+\infty} \longrightarrow \bar{x} \in \bigcap_j P_j = M;$$

where M is the set of solutions of the system; x_0 is an arbitrary initial element of the sequence $\{x_k\}$ generated by the operator $\varphi(x)$.

39. FEJER MAPPINGS AND THEIR PROPERTIES

We suppose that $\varphi \in \{\mathbb{R}^n \to \mathbb{R}^n\}$, $M(\varphi) = \{x \mid \varphi(x) = x\}$, i.e., $M(\varphi)$ is the immobility set of the mapping φ.

Definition 39.1. A mapping φ is called *M-Fejer* if

$$\varphi(y) = y, \quad \forall y \in M;$$
$$\|\varphi(x) - y\| < \|x - y\|, \quad \forall y \in M, \quad \forall x \notin M.$$

Definition 39.2. A sequence $\{x_k\} \subset \mathbb{R}^n$, $\{x_k\} \cap M = \varnothing$ is called *M-Fejer* if

$$\|x_{k+1} - y\| < \|x_k - y\|, \quad \forall k, \quad \forall y \in M.$$

We generalize now Definition 39.1 for the case of many-valued mapping $\varphi(x)$.

Definition 39.3. A many-valued mapping $\varphi(x) \in \{\mathbb{R}^n \to 2^{\mathbb{R}^n}\}$ is called *M-Fejer* if

$$\varphi(y) = y, \quad \forall y \in M;$$
$$\|z - y\| < \|x - y\|, \quad \forall y \in M, \quad \forall x \notin M, \quad \forall z \in \varphi(x).$$

The class of all M-Fejer mappings we denote as F_M.

Below, we shall give some properties of Fejer mappings and Fejer sequences.

Property 39.1. Suppose $\varphi \in F_M$ and the sequence $\{x_k\}$ is reccurently generated by the relation $x_{k+1} \in \varphi(x_k)$, where x_0 is arbitrary. If $\{x_k\} \cap M = \varnothing$, then $\{x_k\}$ is a M-Fejer sequence.

Property 39.2. If $\{x_k\}$ is a M-Fejer sequence, $y \in M$, and $x' \in \{x_k\}'$, then

$$x' \in S_y = \{x \mid \|x - y\| = R_y := \inf_k \|x_k - y\|\}.$$

Here $\{x_k\}'$ is the set of limit points of the sequence $\{x_k\}$. This set $\{x_k\}'$ is, evidently, not empty.

Property 39.3 (follows from the last one). If $x', x'' \in \{x_k\}'$, then

$$M \subset H = \left\{ x \mid \left(x' - x'', x - \frac{x' + x''}{2} \right) = 0 \right\}.$$

This means that if x' and x'' are two different limit points of a *M-Fejer* sequence $\{x_k\}$, then each point $y \in M$, therefore, the whole set M, lies in the hyperplane which consists of the points equidistant from x' and x''.

Corollaries from Property 39.1

1. If M is a solid set, then $\{x_k\} \to x'$, i.e., $\{x_k\}'$ consists of a single point.

2. If $\{x_k\}' \cap M \neq \varnothing$, then $\{x_k\} \to x' \in M$.

Property 39.4. If a set $M \subset \mathbb{R}^n$ admits at least one Fejer mapping, then it is convex and closed.

Proof. Suppose p and q belong to M, $\bar{x} = \alpha p + (1 - \alpha)q$, $\alpha \in (0, 1)$. We must prove that $\bar{x} \in M$. Suppose $\bar{x} \notin M$. Then, by the definition of an M-Fejer mapping, we have

$$\|\varphi(\bar{x}) - p\| < \|\bar{x} - p\|, \qquad \|\varphi(\bar{x}) - q\| < \|\bar{x} - q\|.$$

As the point $\bar{x}$ lies in the segment $[p, q]$, then

$$\|p - \bar{x}\| + \|q - \bar{x}\| = \|p - q\|.$$

Thereby, we obtain the contradiction:

$$\|p - q\| \leq \|\varphi(\bar{x}) - p\| + \|\varphi(\bar{x}) - q\| < \|\bar{x} - p\| + \|\bar{x} - q\| = \|p - q\|.$$

Continuity of the norm $\|\cdot\|$ in $\mathbb{R}^n$ yields completeness of M.   □

Property 39.5. Suppose $\{x_k\}$ is an M-Fejer sequence, and M is convex, closed and solid. Then $\{x_k\} \to x'$.

Proof. This property follows from Property 39.3.   □

These are the simplest properties of Fejer mappings and sequences. We shall use them further without special references.

By the degree $\varphi^t(x)$ of the mapping $\varphi(x)$ is meant subsequent applying of the mapping $\varphi(x)$ t times, i.e., $\varphi^t(x) = \underbrace{\varphi \cdots \varphi}_{t}(x)$.

Lemma 39.1. *If an one-to-one mapping $\varphi \in F_M$ is continuous, then $\{\varphi^t(x_0)\}_{t=1}^{+\infty} \longrightarrow x' \in M$.*

Proof. We must prove that the sequence $\{x_t = \varphi^t(x_0)\}_t$ converges to $x' \in M$ if $\{x_t\} \cap M = \varnothing$. Suppose $x' \in \{x_t\}'$ and $\{x_{j_t}\} \to x'$. We can set that $\{x_{j_t+1}\} \to x''$. If $x' \in M$ then, by Corollary 2 from Property 39.3, it is so. Suppose now that $x' \notin M$; then, as φ is continuous, we have $x'' = \varphi(x')$. Therefore, $\|x'' - y\| < \|x' - y\|$ for $y \in M$, which contradicts Property 39.2. The lemma is proved. □

The lemma holds for pointwise-set mapping $\varphi(x)$ correspondent to Definition 39.3 with the similar proof.

Theorem 39.1. *If a many-valued mapping $\varphi \in F_M$ is closed, i.e., $\{x_k\} \to x'$, $\{y_k\} \to y'$; $y_k \in \varphi(x_k)$ yields $y' \in \varphi(x')$, then the sequence $\{z_t\}$ generated by relation $z_{t+1} \in \varphi(z_t)$ for an arbitrary initial value x_0 converges to $z' \in M$.*

Theorem 39.2. *Suppose $\varphi_j \in F_{M_j}$ and $\bigcap_{j=1}^m M_j =: M \neq \varnothing$. Then*

1. $\varphi(x) := \sum_{j=1}^m \alpha_j \varphi_j(x) \in F_M, \quad \alpha_j > 0, \quad \sum_{j=1}^m \alpha_j = 1;$

2. $\varphi(x) := \varphi_1 \cdots \varphi_m(x) \in F_M.$

Proof. The proof can be done by immediate computation. □

Note that if in p.1 of the theorem all M_j are one and the same set, then the condition for α_j can be weakened: $\alpha_j \geq 0$, $\sum \alpha_j = 1$, which means that α_j may be not strictly positive.

This theorem yields the following corolary.

If mappings $\varphi_j(x)$ are continuous (or closed) then

$$\{\varphi^t(x_0)\}_t \to x' \in M$$

for each written mapping φ. In composition of Fejer mappings the order may be various. In this case, all the arised mappings will be Fejer mappings.

We consider now one general construction of an M-Fejer mapping which can be seen in other contructions. Suppose $d(x)$ is a convex function satisfying the property

$$\{x \mid d(x) \leq 0\} = M,$$

where $\partial d(p)$ is its subdifferential in the point p (see Appendix, Section A2).

Lemma 39.2. *The mapping*

$$x \longrightarrow \left\{ x - \lambda \frac{d^+(x)}{\|h\|^2} \ \Big| \ h \in \partial d(x) \right\} \tag{39.1}$$

is a M-Fejer closed mapping. Here $\lambda \in (0,2)$.

Proof. The property to be Fejer follows from Lemma 40.1 (see Section 40) and from completeness of the mapping $x \to \partial d(x)$. The completeness is verified easily. $\square$

We see that there exist many ways of constructing new M-Fejer mappings in the class F_M using the selected (base) mappings. The algebra in F_M is thus arised which induces new mappings. Using the operations introduced already we may write rather general construction of polynomial type: if $\{\varphi_i(x)\}_1^m \subset F_M$, then

$$\varphi(x) := \sum_{i=1}^m \alpha_i \varphi_{i1}^{n_{i1}} \cdots \varphi_{im}^{n_{im}}(x) \in F_M, \tag{39.2}$$

where $\alpha_i \geq 0$, $\sum_{i=1}^m \alpha_i = 1$. If the mappings $\{\varphi_j(x)\}$ are closed, then $\varphi(x)$ is closed also; therefore, $\{\varphi^t(x_0)\}_t \to x' \in M$.

40.   BASIC CONSTRUCTIONS OF FEJER MAPPINGS FOR ALGEBRAIC POLYHEDRONS

Let φ be a mapping of the space into itself. We consider the two problems. The first problem is to find the set of fixed points $M = \{x \mid x \in \varphi(x)\}$. In particular, we have to find $\bar{x} \in M\}$. The second problem is to construct the mapping $\varphi(x)$ if M is given so that M is the set of fixed points of φ. The both problems will be illustrated below. If, for example, M is a set of solutions of linear inequality system

$$Ax \leq b \sim l_j(x) := (a_j, x) - l_j \leq 0, \qquad j = 1, \ldots, m, \qquad (40.1)$$

then we can set the problem of constructing the M-Fejer mappings in order to construct the iteration methods for solving this system. The example of such M-Fejer mapping is

$$\varphi(x) = x - \frac{1}{m} \sum_{j=1}^{m} \frac{l_j^+(x)}{\|a_j\|^2} a_j.$$

We shall construct below the class of M-Fejer mappings which provide the convergence of appropriate iteration processes.

40.1. Special constructions of Fejer mappings

We suppose here that

$$P := \{x \mid l(x) := (a, x) - b \leq 0\}$$

is a proper half-space from $\mathbb{R}^n$, $\|a\| \neq 0$. The projection of x onto P we define as follows:

$$\varphi(x) = x - \frac{l^+(x)}{\|a\|^2} a.$$

We introduce the *relaxation* mapping $\varphi^\lambda(x)$ correspondent to the mapping $\varphi(x)$:

$$\varphi^\lambda(x) = x - \lambda \frac{l^+(x)}{\|a\|^2} a, \qquad 0 < \lambda < 2.$$

The number λ is called the *relaxation coefficient*. If $\lambda = 2$ the relaxation is called *mirror-like*.

Lemma 40.1. *The mapping $\varphi^\lambda(x)$ is P-Fejer.*

Proof. The proof follows from the trivial identity:

$$\|\varphi^\lambda(x) - y\|^2 = \|x - y\|^2 + 2\lambda \frac{l(x) l(y)}{\|a\|^2} - \lambda(2 - \lambda) \frac{l^2(x)}{\|a\|^2}$$

for $y \in P$, $x \notin P$. □

Lemma 40.2. *The operation* $\varphi_M(x)$ *of projection* x *onto a convex closet set* $M \subset \mathbb{R}^n$ *is* M*-Fejer.*

Proof. We suppose that $p \notin M$, $\bar{p} = \varphi_M(p)$. By Lemma 35.3, the hyperplane $H = \{ x \mid (p - \bar{p}, x - \bar{p}) = 0 \}$ is supporting to M in the point $\bar{p}$; and the half-space $P = \{ x \mid (p - \bar{p}, x - \bar{p}) \leq 0 \}$ contains the set M. The projection of the point p on M will be the projection on P also; therefore, by the inclusion $P \supset M$ and Lemma 40.1, we have

$$\|\varphi_M(p) - y\| = \|\varphi_P(p) - y\| < \|p - y\|, \quad \forall y \in M,$$

which was to be proved. $\square$

The mapping which relaxate the projection on M, namely

$$\varphi(x) = x - \lambda[x - \pi_M(x)], \quad \lambda \in (0, 2),$$

will be M-Fejer also.

For system (40.1) we define

$$\varphi_j(x) = x - \lambda_j \frac{l_j^+(x)}{\|a_j\|^2} a_j, \quad \lambda_j \in (0, 2).$$

Using the mappings $\{\varphi_j(x)\}$ we can construct the mappings

$$\varphi^{(1)}(x) = \sum_{j=1}^{m} \alpha_j \varphi_j(x), \quad \alpha_j > 0, \quad \sum_{j=1}^{m} \alpha_j = 1 \tag{40.2}$$

$$\left(\text{i. e., } \varphi^{(1)}(x) = x - \sum_{j=1}^{m} \lambda_j \alpha_j \frac{l_j^+(x)}{\|a_j\|^2} a_j \right);$$

$$\varphi^{(2)}(x) = \varphi_1 \cdots \varphi_m(x); \tag{40.3}$$

$$\varphi^{(3)}(x) = x - \lambda \frac{l_{j_x}^+(x)}{\|a_{j_x}\|^2} a_{j_x}, \quad \lambda_j \in (0, 2), \tag{40.4}$$

where $j_x \in \{ j \mid \max_{(j)} l_j^+(x) = l_{j_x}(x) \}$.

The mappings $\varphi^{(1)}(x)$ and $\varphi^{(2)}(x)$ are continuous. As for $\varphi^{(3)}(x)$, it is ambiguous (since the choice of j_x is ambiguous) bu it is closed:

$$\{x_t\} \to x', \quad \{y_t\} \to y'; \quad y_t \in \varphi^{(3)}(x_t) \Rightarrow x' \in \varphi^{(3)}(y').$$

This property may be immediately verified.

We shall say that the mappings $\varphi^{(1)}(x)$, $\varphi^{(2)}(x)$, and $\varphi^{(3)}(x)$ are *weighted, subsequent,* and *extremal* projections respectively.

The coefficients in (40.2) may be chosen variously. For example, $\alpha_j = \dfrac{\|a_j\|^2}{\sum_j \|a_j\|^2}$, then we obtain the mapping

$$\varphi^{(4)}(x) = x - \frac{\lambda}{\delta} \sum_{j=1}^{m} l_j^+(x)\, a_j. \tag{40.5}$$

Here $\delta = \sum_{j=1}^{m} \|a_j\|^2$, λ is the general value for λ_j from the formula for $\varphi_j(x)$.

The coefficients α_j in (40.2) may be variable. For example, we can set

$$\alpha_j(x) = \frac{l_j^+(x)}{\delta(x)}, \qquad \delta(x) = \sum_{j=1}^{m} l_j^+(x).$$

This will correspond to the mapping

$$\varphi^{(5)}(x) = x - \frac{\lambda}{\delta(x)} \sum_{j=1}^{m} \frac{l_j^{+2}(x)}{\|a_j\|^2} a_j, \qquad \lambda \in (0,2). \tag{40.6}$$

40.2. General realizations

The construction of types (39.1) and (39.2) we shall call general constructions of Fejer mappings. The sense of this term is as follows: the mapping of type (39.1) are *root mappings* for whole set F_M and (39.2) accumulate in itself each concrete collection of such mappings.

Mapping (39.1) is constructed on the basis of considering the convex function $d(x)$ which specifies the set M. If M is the polyhedron of system (40.1), then $d(x)$ may be any of the sets

$$\max_{(j)} l_j(x), \qquad \sum_{(j)} l_j^+(x), \qquad \|(Ax - b)^+\|^2$$

and so on. According to the choice of $d(x)$ we obtain mapping (39.1). So, Lemma 39.2 is important in the sense of construction of new M-Fejer mappings. Construction (39.2) serves this purpose also.

40.3. The convergence theorem

Theorem 40.1. *If the mapping $\varphi(x)$ from (39.2) is constructed using the collection of mappings $\{\varphi^{(s)}(x)\}_{s=1}^{5}$ give by relations (40.2)–(40.2), then, for each initial x_0, the process $\{x_{t+1} \in \varphi(x_t)\}_t$ converges to solution of compatible system of linear inequalities (40.1), i.e., $\{x_t\}_t \to x' \in M$.*

In formulation of this theorem we may start from any finite collection of M-Fejer mappings satisfying the conditions of Theorem 39.1.

In Section 40.1 we defined the mirror-like case $\lambda = 2$. The following result is interesting ($\lambda = 2$) and we shall formulate it without the proof.

If in (40.2)–(40.2) all λ_j and λ are equal to 2 and the polyhedron of system (40.1) is solid (this corresponds to solvability of the system of strict inequalities $Ax < b$), then each of processes $\{x_{t+1} \in \varphi^{(s)}(x_t)\}_t$, $s = 1, 2, 3$ is teminated in a finite number of steps, i.e. for a certain $\bar{t}$: $x_{\bar{t}} \in M$.

40.4. Convergence rate of Fejer processes

The Fejer processes generated by mappings $\varphi(x) \in F_M$ provide the convergence speed of geometric progression. The geometric progression we characterize by the relation

$$|\varphi(x) - M| \le \Theta|x - M|, \qquad \Theta \in (0, 1).$$

This inequality generates the estimate

$$\|x_t - x'\| \le \Theta^k \|x_0 - x'\|.$$

For the general case, or, for example, for $\varphi(x)$ from Theorem 40.1, the convergence speed cannot be obtained without a certain preliminary work. However, for some particular cases we can obtain the desired estimate on the base of above statements (for example, Lemma 35.5).

Theorem 40.2. *For $\varphi(x)$ given by relation (40.4) the inequality*

$$|\varphi(x) - M| \le \Theta|x - M|, \qquad \Theta \in (0, 1), \tag{40.7}$$

holds for each x.

Proof. For $x \notin M$, we write the inequality from Remark 35.1 to Lemma 35.5

$$|x - M| \le C l_{jx}(x). \tag{40.8}$$

We suppose that $\bar{y} = \pi_M(x)$ so that $l_{jx}(\bar{y}) \le 0$ and

$$|\varphi(x) - M| \le \|\varphi(x) - \bar{y}\|.$$

Hence, taking into account (40.8), we obtain

$$|\varphi(x) - M| \le \left(x - \lambda \frac{l_{jx}(x)}{\|a_{jx}\|^2} a_{jx} - \bar{y}, \ x - \lambda \frac{l_{jx}(x)}{\|a_{jx}\|^2} a_{jx} - \bar{y} \right)$$

$$= |x - M|^2 + 2\lambda \frac{l_{jx}(x)[l_{jx}(\bar{y}) - l_{jx}(x)]}{\|a_{jx}\|^2} + \lambda^2 \frac{l_{jx}^2(x)}{\|a_{jx}\|^2}$$

$$\le |x - M|^2 - \lambda(2 - \lambda) \frac{l_{jx}^2(x)}{\|a_{jx}\|^2} \le |x - M|^2 - \frac{\lambda(2 - \lambda)}{\|a_{jx}\|^2} C^{-2} |x - M|^2$$

$$\le \|x - M\|^2 \left(1 - \frac{\lambda(2 - \lambda)}{C^2 \max_{(j)} \|a_j\|^2} \right) = |x - M|^2 \Theta^2$$

where $\Theta = \left[1 - \dfrac{\lambda(2 - \lambda)}{C^2 \max_{(j)} \|a_j\|^2} \right]^{1/2} \in (0, 1)$. The theorem is proved.   □

The proof of Theorem 40.2 is typical for other settings of M-Fejer mappings also. For example, we consider the mapping $\varphi(x)$ given following (39.1). As the situation is general we confine ourselves by the two conditions.

1) We shall consider inequality (40.7) for $x \in S$ when S is a certain bounded set.

2) A convex function $d(x)$ is majorized from below by the norm $\|(Ax - b)^+\|$ of system residual with a certain constant $\gamma_0 > 0$: $\|(Ax - b)^+\| \le \gamma_0 d(x)$. If these conditions hold, then, from the one hand; we shall have

$$|x - M| \le \gamma_0 C d(x), \tag{40.9}$$

where C is the constant from Lemma 35.5. From the other hand, we shall have

$$\sup_{\substack{h \in \partial d(x) \\ x \in S}} \|h\|^2 = \gamma < +\infty.$$

Then, the similar scheme as when proving Theorem 40.2 leads to inequality (40.7) for $\Theta = \left[1 - \dfrac{\lambda(2 - \lambda)}{C^2 \gamma_0 \gamma} \right]^{1/2} \in (0, 1)$ and $x \in S$. Indeed, supposition of boundedness of S is necessary for geometric speed of convergence for Fejer processes. But, as the sequence $\{x_t\}_t$ induced by mapping $\varphi(x)$ is bounded, we may take, for example, $S = \text{co}\{x_t\}$.

So, if for a M-Fejer mapping $\varphi(x)$, estimate (40.7) holds (this estimate denotes uniform approximation of $\varphi(x)$ extremal relatively M to each point $y \in M$) then, for the points of the process $\{x_{t+1} \in \varphi(x_t)\}_t$ we have the inequality

$$|x_{t+1} - M| \leq \Theta|x_t - M|, \qquad \forall t.$$

This means that $|x_t - M| \leq \Theta^t|x_0 - M|$ which denotes the geometric speed of convergence.

41. DECOMPOSITION AND PARALLELIZING OF FEJER PROCESSES

The structure of Fejer mappings and processes generated by them gives us the possibility of converting them. This is defined, at first, by possibility to construct these iterative mappings via particular mappings which are related to subsystems of the initial system. In this case, the partition of the system on the subsystems may be arbitrary. Secondly, the structure of the matrix of restrictions (block, diagonal, block-diagonal and so on) itself suggests one or another way of constructing the mapping $\varphi \in F_M$.

The scheme of converting is as follows:

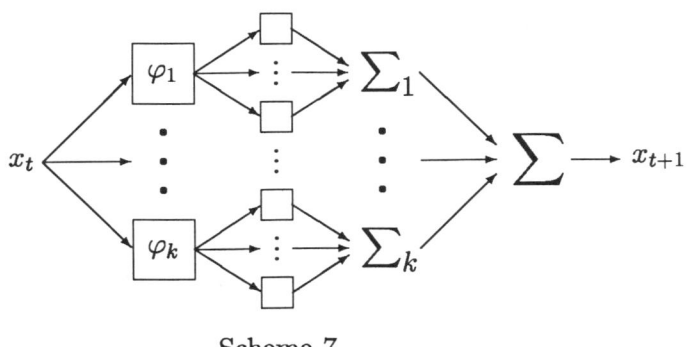

Scheme 7

We suppose that the system $Ax \leq b$ is divided into the subsystems

$$A_j x \leq b_j, \qquad j = 1, \ldots, k \tag{41.1}$$

with the sets of solutions M_j so that

$$M := \{x \mid Ax \leq b\} = \bigcap_{j=1}^{k} M_j.$$

If $\varphi_j \in F_{M_j}$, then the M-Fejer mapping $\varphi \in F_M$ can be constructed, for example, using the scheme (40.2) for $\alpha_j = 1/k$:

$$\varphi(x) = (1/k) \sum_{j=1}^{k} \varphi_j(x).$$

In Scheme 7 this will be correspondent to calculating the elements $x_t^s :=$ $\varphi_s(x_t)$, $s = 1, \ldots, k$, which are sended on the summator $\sum$ with obtaining arithmetic mean $(1/k) \sum_{j=1}^{k} x_t^s = x_{t+1}$. Derivation of the elements x_t^s may be converting also, for example, at the expense of converting the calculations of scalar products (x_t, a_j) and the norms $\|a_j\|$, $j = 1, \ldots, m$. Such possibility is just foreseen in Scheme 7.

We shall give now the example of realization of this scheme to solving the LP problem:

$$L: \ \max\{(c, x) \mid Ax \leq b\}.$$

The dual problem to this one will be

$$L^*: \ \min\{(b, u) \mid A^T u = c, \ u \geq 0\}.$$

This pair of mutually dual problems is reduced to the system

$$S: \ \begin{cases} Ax \leq b, \\ \qquad\qquad A^T u = c, \qquad u \geq 0, \\ (c, x) - (b, u) \geq 0 \end{cases}$$

(see (15.3) and Theorem 15.5). The last inequality in the system S may be replaced by the equation $(c, x) = (b, u)$. The matrix of the system S has the block form

$$\begin{bmatrix} A & O \\ O & A^T \\ c^T & -b^T \end{bmatrix}$$

Accordingly this block structure we can select its subsystems with the polyhedrons

$$M = \{x \mid Ax \leq b\}, \qquad M^* = \{u \geq 0 \mid A^T u = c\}$$

and the half-space

$$P = \left\{ \begin{bmatrix} x \\ u \end{bmatrix} \;\middle|\; (c, x) \geq (b, u) \right\}.$$

If $\varphi \in F_M$, $\varphi^* \in F_{M^*}$ and $\varphi_0(z) = \pi_P(z)$, $z = \begin{bmatrix} x \\ u \end{bmatrix}$, then converting the Fejer process for the system S (therefore, simultaneous solving the problem L and L^*) can be based on the following scheme

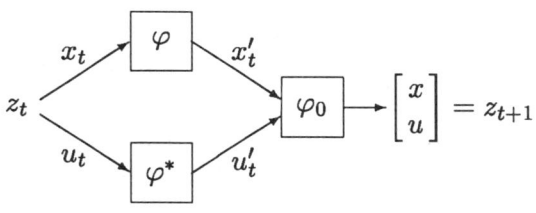

Scheme 8

Realization of this scheme is analytic form will be as follows : if $z_t = \begin{bmatrix} x_t \\ u_t \end{bmatrix}$, then we derive in a parallel way $x'_t = \varphi(x_t)$, $u'_t = \varphi^*(u_t)$ and then $z_{t+1} = \varphi_0^+(z'_t)$, $z' = \begin{bmatrix} x'_t \\ u'_t \end{bmatrix}$. Here φ_0^+ we denote the positive cut with respect to the second component of the vector $\varphi_0(z'_t)$. The result Fejer operator generated the sequence $\{z_t\}_t$ can be written in the following compact form

$$\Psi(z) := \varphi_0^+(\varphi(x), \varphi^*(u)).$$

The system S can be reduced to the system with partially non-negative terms

$$Ax + v = b, \quad A^T u = c, \quad (c, x) - (b, u) = 0, \quad [v, u] \geq 0.$$

In this connection, and following the general interest, we consider the converting of the Fejer process for the system

$$Ax = b, \qquad x \geq 0. \tag{41.2}$$

This representation is not connected with the sense of A or b in the previous representations (inequalities systems, LP problem and so on). We suppose that $x \in \mathbb{R}^n$ and the rows of the matrix A are linearly independent. Then the matrix AA^T of $m \times m$ dimension will be regular; therefore, $(AA^T)^{-1}$ will exist. If $H = \{x \mid Ax = b\}$, then, when constructing the M-Fejer mapping for $M = H \cap \mathbb{R}^n_+$, we can use the known operator of projection on H:

$$\pi_H(x) = x - A^T(AA^T)^{-1}(b - Ax). \tag{41.3}$$

It will be H-Fejer, and $\varphi(x) = \pi_H^+(x)$ will be M-Fejer. As $\varphi(x)$ is continuous, then

$$\{\varphi^t(x_0)\}_t \to x' \in M,$$

which means that the process generated by $\varphi(x)$ converges to the solution of system (41.2). If system (41.2) would have the form

$$Ax \leq b, \qquad x \geq 0, \tag{41.4}$$

then, writing it in the form

$$Ax + v = b, \qquad x \geq 0, \qquad v \geq 0, \tag{41.5}$$

we can apply the above construction. As a result we obtain

$$\varphi(z) = [z - \bar{A}^T(\bar{A}\bar{A}^T)^{-1}(b - \bar{A}z)]^+, \tag{41.6}$$

where $z = \begin{bmatrix} x \\ v \end{bmatrix}$, $\bar{A} = \begin{bmatrix} A & \begin{matrix} 1 & & \\ & \ddots & \\ & & 1 \end{matrix} \end{bmatrix}$. If $\{\varphi^t(z_0)\} \to \begin{bmatrix} x' \\ v' \end{bmatrix}$, then x' solves system (41.4). Note that in (41.4) it is not necessary that the rows of A linearly independent, and in system (41.5) this holds automatically. This provides the existence of the inverse problem $(\bar{A}\bar{A}^T)^{-1}$ in formula (41.6).

Converting calculation of $\varphi(x)$ for system (41.2) with applying the operator of projection on the linear manifold can be done on the basic of partition of type (41.1). Namely, we take an arbitrary partition

$$A_j x = b_j, \qquad j = 1, \ldots, k \qquad (41.7)$$

of system (41.2). We consider the operators

$$\varphi_j(x) = x - A_j^T (A_j A_j^T)^{-1} (b_j - A_j x).$$

These are the operators of projection on the manifolds

$$H_j = \{x \mid A_j x = b_j\}, \qquad j = 1, \ldots, k.$$

From $\{\varphi_j(x)\}$ we can form $\varphi \in F_M$, $M = \{x \geq 0 \mid Ax = b\}$ in various ways, for example:

$$\varphi(x) = \left[\sum_{j=1}^{k} \alpha_j \varphi_j(x) \right]^+ ,$$

where $\alpha_j > 0$, $\sum_{j=1}^{k} \alpha_j = 1$. Converting the process $\{\varphi^t(x_0)\}_t$ on the basis of partition of initial system into subsystems corresponds to Scheme 7.

Finally, we realize the converting Scheme 7 responded the partition of the restriction system into subsystems both in direct and in dual problems in the form:

$$L: \quad \max\{(c, x) \mid Ax \leq b, \ x \geq 0\},$$

$$L^*: \quad \min\{(b, u) \mid A^T u \geq c, \ u \geq 0\}.$$

To this end we rewrite them as follows:

$$\max\{(c, x) \mid Ax + v = b, \ x \geq 0, \ v \geq 0\},$$

$$\min\{(b, u) \mid A^T u - w = c, \ u \geq 0, \ v \geq 0\}.$$

Reduction $\{L, L^*\} \xrightarrow{\sim} S$ in this case leads to the system

$$\begin{cases} Ax + v = b, \quad A^T u - w = c, \\ \qquad (c, x) = (b, u), \\ x \geq 0, \quad v \geq 0, \quad u \geq 0, \quad w \geq 0. \end{cases} \qquad (41.8)$$

We divide the matrix A into horizontal A_j, $j = 1, \ldots, k$ and vertical H_i, $i = 1, \ldots, l$ submatrices arbitrarily. The intersection of A_j and H_i we denote by A_{ji}. The correspondent partition of the vectors c, x, w and b, u, v into the subvectors we denote as c_i, x_i, w_i and b_j, u_j, v_j respectively. In these notations system (41.8) will be written as follows:

$$\sum_{i=1}^{l} A_{ji} x_i + v_j = b_j, \tag{41.9_1}$$

$$\sum_{j=1}^{k} A_{ji}^T u_j - w_i = c_i, \tag{41.9_2}$$

$$\sum_{i=1}^{l} (c_i, x_i) = \sum_{j=1}^{k} (b_j, u_j), \tag{41.9_3}$$

$$x_i \geq 0, \quad v_j \geq 0; \quad u_j \geq 0, \quad w_i \geq 0. \tag{41.9_4}$$

The multifold H'_j, $j = 1, \ldots, k$ is the set of solutions of system (41.9_1) and H''_i, $i = 1, \ldots, l$ is the set of solutions of system (41.9_2). We denote by H_0 the hyperplane with equation (41.9_3) and select (constructively) the mappings

$$\varphi_j \in F_{H'_j}, \qquad \varphi_j^* \in F_{H''_j}, \qquad \varphi_0 \in F_{H_0}. \tag{41.10}$$

These mappings may be, in particular, the projections on the correspondent multifolds. It is important that when we program operators (41.10) we remain the possibility of converting the calculations with respect to the introduced partition of matrices and vectors. This means that the elements of converting certain operations, for example: $(a_j, x) = \sum_{i=1}^{l}(a_j^i, x)$, $\|a_j\|^2 = \sum_{i=1}^{l}(x_i, x_i)$ and so on are inserted into numerical realization of operators (41.10). The conjunction of all mappings (41.10) into only one we may do, for example, by the formula

$$\Psi(x, u, v, w) := \pi_{H_0}^+ \left(\sum_{j=1}^{k} \alpha_j \varphi_j(x, v), \ \sum_{i=1}^{l} \beta_i \varphi_i^*(u, w) \right).$$

Here $\alpha_j > 0$, $\sum_{j=1}^{k} \alpha_j = 1$, $\beta_i > 0$, $\sum_{i=1}^{l} \beta_i = 1$; "+" in $\pi_{H_0}^+$ denotes the cut by the whole system of variables (41.9_4). The operator Ψ is continuous (if operators from (41.10) are continuous) and $\Psi \in F_M$, where M is the solution set of system (41.8). Therefore,

$$\{\Psi^t(z)\}_t \longrightarrow z' = \begin{bmatrix} x' \\ u' \\ v' \\ w' \end{bmatrix}$$

where $x' \in \operatorname{Arg} L$, $u' \in \operatorname{Arg} L^*$.

42. RANDOMIZATION OF FEJER PROCESSES

Fejer mappings being the iterative operators for systems of linear inequalities and LP problems are very convenient for their combined employment. We can see this from the results of previous paragraph. Namely, if $\{\varphi_j(x)\}_1^m \subset F_M$ is a finite number of M-Fejer mappings such that

$$\{\varphi_j^t(x_0)\}_t \longrightarrow x' \in M,$$

then we may use them in a certain iterative process following a certain mixed strategy

$$x_{t+1} = \begin{cases} \varphi_1(x_t) & \text{—with probability } P_1, \\ \cdots\cdots\cdots\cdots\cdots\cdots\cdots\cdots\cdots \\ \varphi_m(x_t) & \text{—with probability } P_m. \end{cases} \tag{42.1}$$

Here $P_j \geq 0$, $\sum_{j=1}^m P_j = 1$. The vector $P = [P_1, \ldots, P_m]$ we shall call the *mixed strategy* with the use of operators $\varphi_j(x)$ for generating the iterative sequence. The mixed strategy can be changed in each step t: $P^t = [P_1^t, \ldots, P_m^t]$. The evolution of strategy P^t may be defined relatively each φ_j, by diminishing a certain residual function $d(x)$ for the set M. If M is the polyhedron of the system $l_j(x) \leq 0$, $j = 1, \ldots, m$, then $d(x)$ may be any of the function $\max_{(j)} l_j^+(x)$, $\sum_{(j)} l_j^+(x)$, and so on.

The operator attaining its value respectively (42.1) we denote by $\varphi^P(x)$. This operator is realized on a certain $\varphi_{j_P}(x)$. If $P = P^t$, we shall write $\varphi^{P^t}(x) =: \varphi_{j_t}(x)$.

Definition 42.1. An operator $\varphi \in F_M$ we shall call *regular* if it satisfies relation (40.7).

Theorem 42.1. *Suppose the following conditions hold*

1. $\{\varphi_j(x)\}_1^m \subset F_M$ *and at least one operator from this collection is either regular or continuous.*

2. *Coordinates of the sequence $\{P^t\}$ of mixed strategies are uniformly separated from zero (i.e. $\exists \varepsilon > 0: P_j^t \geq \varepsilon, \forall j, \forall t$).*

Then the sequence $\{x_t\}$ setted by the recurrent relation

$$x_{t+1} = \varphi^{P^t}(x_t) \quad (= \varphi_{j_t}(x_t)) \tag{42.2}$$

converges to $x' \in M$.

Proof. We suppose first that one of the operators $\{\varphi_j(x)\}_1^m$ is regular, for example, $\varphi_1(x)$. From the condition $P_1^t \geq \varepsilon > 0$ it follows that there exists a sequence $\{x_{t_k}\} \to x'$ such that $x_{t_k+1} = \varphi_1(x_{t_k}) \to x''$. If $x' \in M$, then $\{x_t\} \to x'$ (follows from Property 39.3). If $x' \notin M$, then, by the Condition 1, we have

$$|x_{t_k+1} - M| = |\varphi_1(x_{t_k} - M| \leq \Theta |x_{t_k} - M|.$$

Hence it follows that

$$|x'' - M| \leq \Theta |x' - M| < |x' - M|.$$

Evidently, there exists a vector $y \in M$ such that $\|x'' - y\| < \|x' - y\|$ which contradicts Property 39.2.

If, for example, $\varphi_1(x)$ is continuous, then, as in the previous case, $\{x_{t_k}\} \to x'$ yields $x_{t_k+1} = \varphi_1(x_{t_k}) \to x''$. Passing to the limit in the inequality

$$\|\varphi_1(x_{t_k}) - y\| \leq \|x_{t_k} - y\|, \qquad y \in M,$$

we obtain

$$\|x'' - y\| = \|\varphi_1(x') - y\| \leq \|x' - y\|.$$

In this connection, if $x' \notin M$, then we obtain the strict inequality $\|x'' - y\| < \|x' - y\|$ which contradicts to Property 39.2. The theorem is proved. $\quad\square$

As was above noted, the recount of the mixed strategy P^t on each step can be done depending on the contribution of each $\varphi_j(x)$ in diminishing the residual function $d(x)$. We show now several ways of recounting P^t.

We suppose that $\varphi_j(x_t) = x_t^j$, $\delta_t^j = [d(x_t) - d(x_t^j)]^+$. Here we have used the cut since monotone decreasing of the residual function is not guaranteed in Fejer processes. We set

$$P_j^{t+1} = \frac{\delta_t^j}{\sum_{j=1}^m \delta_t^j}. \tag{42.3}$$

Forming P^{t+1} with coordinates P_j^{t+1}, can be done starting from $\{x_k, \ldots, x_t\}$ setting in (42.3)

$$\delta_t^j = \sum_{s=k}^t [d(x_s) - d(x_s^j)]^+.$$

The numerical experiments were done following this method of control for mixed strategies in Fejer processes. As a rule, beginning with a certain t', all the mappings were rejected except one, i.e.

$$\varphi^{P^t}(x_t) = \varphi_{j'}(x_t), \qquad t \geq t'.$$

In this case there is a good reason to believe that the operator $\varphi_{j'}(x)$ was the most effective.

bf Remark. The additional tests for intermediate estimates of each $\varphi_j(x)$ have their own calculating cost with whoch we must reckon. However, if we take into account the low price of calculating the values $\varphi_j(x)$ and $d(x)$ (in the class of iteration mappings in question), then, for small m (for example $2 \leq m \leq 4$) this way of selection is well founded.

Construction of the trajectory $\{P^t\}$ of mixed strategies (for the Fejer process) can be realized, on the basis of the Brown–Robinson approach (see (Brown, 1951), (Robinson, 1951), (Braithwaite, 1959)) approach suggested by these authors for solving the matrix games in mixed strategies. In application to our situation the recount algorithm is as follows: if P^t is realized on the step t and $\varphi^{P^t}(x_t) = \varphi_{j_{t+1}}(x_t)$, then P^{t+1} is derived by the formula

$$P^{t+1} = \frac{t}{t+1} P^t + \frac{1}{t+1} e_{j_{t+1}} \quad \left(= \frac{1}{t+1} \sum_{s=1}^{t+1} e_{j_s} \right).$$

Here $e_{j_{t+1}} = [0, \ldots, 1, \ldots, 0]$ is the unit vector of the space $\mathbb{R}^n$ with unity in the place j_{t+1} which corresponds to choosing the pure strategy following the relation $\varphi^{P^t}(x_t) = \varphi_{j_{t+1}}(x_t)$. Thereby, the idea of conbined way of the use of certain-iterative mappings is rather fruitful.

43. FEJER PROCESSES AND INCONSISTENT SYSTEMS OF LINEAR INEQUALITIES

Some of M-Fejer mappings (for example, (41.5)) generate converging sequence independently of emptiness or nonemptiness of the set $M = \{x \mid Ax \leq b\}$, i.e., independently of compatibility or incompatibility of the inequality system in question. If the mapping $\varphi \in F_M$ is continuous and $M = \varnothing$, where $\{\varphi^t(x_0)\}_t \to x'$, then, evidently, $x' = \varphi(x')$. Therefore, in this case, the immobility set $\widetilde{M}$ for $\varphi(x)$ is not empty. It is natural to suppose that $\widetilde{M}$ is a certain approximate set for incompatible system of inequalities. Really, it is so.

Fejer mappings with this property can be used for correction problems for incompatible systems of linear inequalities and for $IPLP$ problems.

43.1. Nonexpanding M-Fejer mappings

Definition 43.1. A mapping $\varphi \in \{\mathbb{R}^n \to \mathbb{R}^n\}$ is called *nonexpanding* if $\|\varphi(x) - \varphi(y)\| \leq \|x - y\|$ for all $x, y \in \mathbb{R}^n$.

Definition 43.2. A mapping $\varphi \in \{\mathbb{R}^n \to \mathbb{R}^n\}$ is called *weakly M-Fejer* if

$$\|\varphi(x) - y\| \leq \|x - y\|, \quad \forall y \in M; \quad \varphi(x) \neq x, \quad \forall x \notin M;$$
$$\varphi(y) = y, \quad \forall y \in M.$$

Lemma 43.1. *If $\varphi(x)$ is weakly M-Fejer, then for each $\alpha \in (0,1)$, we have*

$$\varphi_\alpha(x) := (1 - \alpha)\varphi(x) + \alpha x \in F_M.$$

Proof. The proof follows from the fact that $\varphi_\alpha(x)$ lies inside the ball with the center in the point y and of the radius $\|x - y\|$. □

Corollary 43.1. *If* φ *is a nonexpanding mapping and* $M = \{x \mid \varphi(x) = x\} \neq \varnothing$, *then* $\varphi_\alpha(x) := (1 - \alpha)\varphi(x) + \alpha x \in F_M$.

Lemma 43.2. *The operator of projection on a convex set* $M \subset \mathbb{R}^n$ *is a nonexpanding operator, i.e.,*

$$\|\pi_M(p) - \pi_M(q)\| \leq \|p - q\| \tag{43.1}$$

for each $x, y \in \mathbb{R}^n$.

Proof. We set $\pi_M(p) = \bar{p}$, $\pi_M(q) = \bar{q}$. By Lemma 35.3 we have

$$\{x \mid (p - \bar{p}, x - \bar{p}) \leq 0\} \supset M \supset \bar{q};$$

therefore,

$$(p - \bar{p}, \bar{q} - \bar{p}) \leq 0. \tag{43.2}$$

Analogously, we have

$$(q - \bar{q}, \bar{p} - \bar{q}) \leq 0. \tag{43.3}$$

Transforming (43.2) and (43.3) into inequalities

$$(p - \bar{q}, \bar{q} - \bar{p}) + \|\bar{p} - \bar{q}\|^2 \leq 0,$$
$$(q - \bar{p}, \bar{p} - \bar{q}) + \|\bar{p} - \bar{q}\|^2 \leq 0$$

and summing them we obtain

$$2\|\bar{p} - \bar{q}\|^2 \leq (p - q, \bar{q} - \bar{p}) + \|\bar{p} - \bar{q}\|^2.$$

Hence we have $\|\bar{p} - \bar{q}\|^2 \leq \|p - q\| \cdot \|\bar{p} - \bar{q}\|$; i.e., (43.1) holds. □

Lemma 43.3. *The mapping*

$$\varphi^\lambda(x) = x - \lambda \frac{l^+(x)}{\|a\|^2}\, a, \quad \lambda \in (0, 2)$$

is nonexpanding, where $l(x) = (a, x) - b$, $\|a\| \neq 0$.

Proof. If $l(x) \leq 0$ and $l(y) \leq 0$ then $\varphi^\lambda(x) = x$, $\varphi^\lambda(y) = y$; therefore, the required inequality holds. If $l(x) > 0$, $l(y) \leq 0$, then

$$\|\varphi^\lambda(x) - \varphi^\lambda(y)\| = \|\varphi^\lambda(x) - y\| < \|x - y\|,$$

since $\varphi^\lambda(x) \in F_P$ for $P = \{x \mid l(x) \leq 0\}$ (Lemma 40.1).

The case $l(x) \leq 0$, $l(y) > 0$ is considered analogously. We suppose now that $l(x) > 0$, $l(y) > 0$. Then the required inequality yields from the trivial identity

$$\|\varphi^\lambda(x) - \varphi^\lambda(y)\|^2 = \|x - y\|^2 - \frac{\lambda(2 - \lambda)}{\|a\|^2}[l(x) - l(y)]^2.$$

The lemma is proved.   □

Corollary 43.2. *Mappings* (40.2), (40.3), *and* (40.5) *are nonexpanding.*

Proof. Mappings (40.2) and (40.5) are convex combinations on nonexpanding mappings of the type $\varphi^\lambda(x)$ from Lemma 40.1; therefore, they are nonexpanding also. As for mapping (40.3), we have

$$\|\varphi^{(2)}(x) - \varphi^{(2)}(y)\| \leq \|\varphi_2 \cdots \varphi_m(x) - \varphi_2 \cdots \varphi_m(y)\|$$

$$\leq \|\varphi_m(x) - \varphi_m(y)\| \leq \|x - y\|$$

which was to be proved.   □

Lemma 43.4. *Mapping* (40.5), *i.e.,*

$$\varphi(x) = x - \frac{\lambda}{\delta} \sum_{j=1}^{m} l_j^+(x)\, a_j, \quad \lambda \in (0, 2), \quad \delta = \sum_{j=1}^{m} \|a_j\|^2, \tag{43.4}$$

is a Fejer mapping relatively the set

$$\widetilde{M} = \text{Arg min} \sum_{j=1}^{m} l_j^+(x).$$

Proof. We write (43.4) in the form

$$\varphi(x) = x - \frac{\lambda}{2\delta} \nabla \sum_{j=1}^{m} l_j^{+2}(x). \tag{43.5}$$

We see that $\widetilde{M} = \{x \mid \varphi(x) = x\}$, i.e., $\widetilde{M}$ is the set of fixed points for the mapping $\varphi(x)$. We take such $\alpha \in (0,1)$ that $\alpha^{-1}\lambda = \lambda' \in (0,2)$. Then $\varphi(x)$ can be rewritten in the form

$$\varphi(x) = \alpha \varphi^{\lambda'}(x) + (1 - \alpha)x,$$

where

$$\varphi^{\lambda'}(x) = x - \frac{\lambda'}{\delta} \sum_{j=1}^{m} l_j^+(x)\, a_j.$$

By Lemma 43.3 and Corollary 43.1, the mapping $\varphi^{\lambda'}(x)$ is nonexpanding. Then, applying Lemma 43.1, we obtain that $\varphi \in F_{\widetilde{M}}$, which was to be proved. $\square$

Lemma 43.5. *Mapping* (40.2)

$$\varphi(x) = x - \sum_{j=1}^{m} \alpha_j \lambda_j \frac{l_j^+(x)}{\|a_j\|^2}\, a_j \tag{43.6}$$

where $\lambda_j \in (0,2)$, $\alpha_j > 0$, $\sum_{j=1}^{m} \alpha_j = 1$ *is Fejer relatively the set*

$$\widetilde{M} = \text{Arg min}_{(x)} \sum_{j=1}^{m} \alpha_j \lambda_j \frac{l_j^{+2}(x)}{\|a_j\|^2}.$$

Proof. The proof is analogous to the previous one. $\square$

43.2. The Fejer process for the problem of square approximation of an incompatible system of linear inequalities

Theorem 43.1. *The process $\{\varphi^t(x_0)\}_t$ for $\varphi(x)$ of form (43.4) or (43.6) converges to $x' \in \text{Arg min}_{(x)}\, d(x)$, where $d(x) = \|(Ax - b)^+\|^2$ and $d(x) = \sum_{j=1}^m \alpha_j \lambda_j \frac{l_j^{+2}(x)}{\|a_j\|^2}$ respectively.*

Proof. The proof follows from continuity of mappings (43.4), (43.6) and $\widetilde{M}$-Fejer property of their immobility sets. $\square$

We shall give nopw generalization of the mappings $\varphi(x)$ from Theorem 43.1 related to the mixed system of linear inequalities and equations

$$\begin{aligned} l_j(x) &= 0, & j \in J_=, \\ l_j(x) &\leq 0, & j \in J_\leq \end{aligned} \tag{43.7}$$

where $l_j(x) = (a_j, x) - b_j$, $j \in J_= \cup J_\leq$. The function of square (more precisely, piecewise square) residual for this system will be as follows:

$$d(x) = \sum_{J_=} l_j^2(x) + \sum_{J_\leq} l_j^{+2}(x). \tag{43.8}$$

The problem $\min d(x)$ is the problem of square approximation for system (43.7). It is important by itself; besides, *LP* problem, regularized *LP* problems, problems of square programming, and others are reduced to this problem. For (43.7) we write the analog of (43.4)

$$\varphi(x) = x - \frac{\lambda}{\delta}\left[\sum_{J_=} l_j(x)\,a_j + \sum_{J_\leq} l_j^+(x)\,a_j\right] \quad \left(= x - \frac{\lambda}{2\delta}\nabla d(x)\right); \tag{43.9}$$

here $\delta = \sum_J \|a_j\|^2$, $J = J_= \cup J_\leq$.

We suppose that $\widetilde{M} = \text{Arg min}\, d(x)$. Analogously to Lemma 43.4 we can prove the following lemma.

Lemma 43.6. *Mapping (43.9) is continuous and $\widetilde{M}$-Fejer. Hence it follows:*

$$\{\varphi^t(x_0)\} \longrightarrow x' \in \widetilde{M}.$$

43.3. Extension of the results to the case of system (43.7) with additional condition $x \in S$

We consider system (43.7) with the condition $x \in S$, where S is a certain convex closed set. The problem is as follows:

$$\min_{x \in S} d(x), \qquad (43.10)$$

where $d(x)$ is a convex differentiable function. For example, function (43.8) is such a function.

Lemma 43.7. *The vector $\bar{x} \in S$ is optimal for solvable problem (43.10) if and only if*

$$\pi_S(\bar{x} - \gamma \nabla d(\bar{x})) = \bar{x}, \quad \gamma > 0. \qquad (43.11)$$

Proof. The case $\nabla d(\bar{x}) = 0$ is trivial. In this case $\bar{x}$ is the absolute minimum of $d(x)$. Suppose now that $\nabla d(\bar{x}) \neq 0$. Let $\bar{x} \in \widetilde{S} := \mathrm{Arg}\,(43.10)$. The hyperplane

$$H = \{x \mid (\nabla d(\bar{x}),\, x - \bar{x}) = 0\}$$

will be plane of support in the point $\bar{x}$. Besides, $(\nabla d(\bar{x}),\, x - \bar{x}) \geq 0,\, \forall x \in S$. The point $\bar{z} = \bar{x} - \gamma \nabla d(\bar{x})$ has one and the same projection on S and on H. It is the point $\bar{x}$, which corresponds (43.11). Conversely, suppose (43.11) holds. We have the hyperplane of support H and the inequality $(\nabla d(\bar{x}),\, x - \bar{x}) \geq 0$ for all $x \in S$. As the function $d(x)$ is convex, then, by Theorem 3 (Appendix, Section A1), we have

$$(\nabla d(\bar{x}),\, x - \bar{x}) \leq d(x) - d(\bar{x})$$

for all x. These two inequalities yield $d(x) \geq d(\bar{x}),\, \forall x \in S$; i.e. $\bar{x} \in \widetilde{S}$. The lemma is proved. □

Suppose now that $\varphi(x)$ is operator (43.9). We define

$$\psi_\alpha(x) = (1 - \alpha)\, \pi_S[\varphi(x)] + \alpha x, \quad \alpha \in (0, 1).$$

Theorem 43.2. *The mapping $\psi_\alpha(x)$ is continuous and $\widetilde{S}$-Fejer, where $\widetilde{S} = \mathrm{Arg}\,(43.10)$. Hence it follows that*

$$\{\psi_\alpha^t(x_0)\}_t \longrightarrow x' \in \widetilde{S}.$$

Proof. The operator $\psi(x) = \pi_S[\varphi(x)]$ is nonexpanding (by Lemmas 43.2 and 43.6). By Lemma 43.7, $\widetilde{M}$ is the immobility set for $\psi(x)$. Then, Lemma 43.1 yields $\psi_\alpha \in F_{\widetilde{S}}$. Continuity of $\psi_\alpha(x)$ follows from continuity of the projection operator $\pi_S(x)$ and continuity of the gradient $\nabla d(x)$ of differentiable convex function $d(x)$ (Theorem 2 from Appendix, Section A1). □

44. FEJER PROCESSES FOR REGULARIZED LP PROBLEMS

We consider the problem

$$L: \quad \max\{(c, x) \mid Ax \le b\}$$

together with the dual problem

$$L^*: \quad \min\{(b, u) \mid A^T u = c,\ u \le 0\}$$

with their admissible sets M and M^*. We shall consider the regularized Fejer processes for L both in the case of its solvability and unsolvability.

44.1. The case when the problem L is solvable

Together with the problem L we consider the regularized problem

$$\max\{(c, x) - \alpha\|x - p\|^2 \mid Ax \le b\}, \quad \alpha > 0. \tag{44.1}$$

By Lemma 38.1, for $\alpha > 0$ sufficiently small; we have

$$x_\alpha := \arg(44.1) = \arg\min\{\|x - p\| \mid x \in \mathrm{Arg}\,L\}. \tag{44.2}$$

Problem (44.1) is transformed as follows: (see Section 38.1)

$$\min\left\{\left\|x - \left(p + \frac{c}{2\alpha}\right)\right\|^2 \;\middle|\; Ax \le b\right\}. \tag{44.3}$$

By the method of square penalty functions (see Theorem 17 from Appendix, Section A6), problem (44.3) is reduced to the problem asymptotically equivalent to the initial one:

$$\min\left\{\left\|x - \left(p + \frac{c}{2\alpha}\right)\right\|^2 + R\|(Ax - b)^+\|^2\right\}, \quad R > 0. \tag{44.4}$$

Let $d_R(x)$ be the function we minimize in (44.4). Then the problem will be written briefly

$$\min_{(x)} d_R(x). \tag{44.5}$$

The vector $x_R \in \mathrm{Arg}\,(44.5)$ will solve problem (44.1) approximately if (44.2) holds. This means that x_R solves the problem of p-normalization of optimal vector of the problem L approximately. The reduction of the problem L to (44.5) solves the problem of its stability and well-posedness. The total result we formulate as the theorem.

Theorem 44.1. *If the problem L is solvable and $\alpha > 0$ is chosen so that (44.2) holds, then*

1) $\{x_R\}$, $x_R \in \mathrm{Arg}\,(44.5)$ *the convergence* $\{x_R\} \to x_\alpha = \arg\min\{\|x - p\| \mid x \in \mathrm{Arg}\,L\}$ *holds.*

2) *if* $\varphi(x)$ *is the mapping constructed as (43.9) for (44.5), then*

$$\{\varphi^t(x_0)\}_t \to x_R \in \mathrm{Arg}\,(44.5).$$

Explanations to the proof. The choice of $\alpha > 0$ which provides (44.2) is possible by Lemma 38.1. By Theorem 17 from Appendix, Section A6, we have

$$\|x_R - x_\alpha\| \leq \frac{\|\bar{u}\|}{2\sqrt{2\alpha R}} \to 0 \quad \text{for} \quad R \to +\infty,$$

which provides the desired convergence. As for the second part of the theorem, the convergence follows from the general theorem on convergence of the process generated by a continuous Fejer operator.

All we have said relatively iterative process for regularized problem (44.1) may be repeated for $\varphi(x)$ constructed following (43.6).

44.2. The case when the problem L is unsolvable

For problem (44.1) the dual problem will be as follows:

$$\min\{F(x,u) \mid \nabla_x F(x,u) = 0, \ u \geq 0\}. \tag{44.1*}$$

(see Appendix, Section A3), where

$$F(x,u) = (c,x) - \alpha \|x - p\| - (Ax - b, u)$$

is the Lagrange function correspondent to problem (44.1). Relations $\nabla_x F(x,u) = 0$, $u = 0$ are rewritten in the form

$$c - 2\alpha(x - p) = A^T u, \quad u \geq 0. \tag{44.6}$$

We suppose that $M = \{x \mid Ax \leq b\}$, $M^* = \{[\bar{x}, u] \mid (44.6)\}$. The set M may be empty and not empty. As for M^*, we have $M^* \neq \varnothing$. Really, for each $\bar{u} \geq 0$ from (44.6), we can take $\bar{x} = p + \dfrac{c - A^T \bar{u}}{2\alpha}$; therefore, $[\bar{x}, \bar{u}] \in M^*$. Following the classification of improper LP problems, for (44.1), the two variants are possible:

1) solvability; this is the case when $M \neq \varnothing$;
2) unsolvability of the first order: $M = \varnothing$, $M^* \neq \varnothing$.

In particular, this denotes that if

$$M_{\Delta b} = \{x \mid Ax \leq b + \Delta b\} \neq \varnothing,$$

then the problem

$$\max\{(c,x) - \alpha \|x - p\|^2 \mid Ax \leq b + \Delta b\} \tag{44.7}$$

is solvable. Note that the regularization of the problem L in form (44.1) makes impossible for the regularized problem to be improper of 2-nd or 3-d orders. Therefore, its optimal correction can be realized at the sacrifice of the increment Δb. The problem of correction we naturally set in the form

$$\max\{(c,x) - \alpha \|x - p\|^2 - R\|\Delta b\|^2 \mid Ax \leq b + \Delta b, \ \Delta b \geq 0\}.$$

The last problem can be rewritten as follows:

$$\max\{(c,x) - \alpha \|x - p\|^2 - R\|(Ax - b)^+\|^2\},$$

or, with an accuracy of a multiplier α,

$$\min\left\{\left\|x - \left(p + \frac{c}{2\alpha}\right)\right\|^2 + \frac{R}{\alpha}\|(Ax - b)^+\|^2\right\}. \tag{44.8}$$

Therefore, the correction problem is rewritten in the form (44.4). If we construct the mapping $\varphi(x)$ following type (43.9), we shall have the convergence

$$\{\varphi^t(x_0)\}_t \to x' \in \widetilde{M} := \{x \mid \nabla d(x) = 0\},$$

where

$$d(x) = \left\| x - \left(p + \frac{c}{2\alpha} \right) \right\|^2 + \frac{R}{\alpha} \|(Ax - b)^+\|^2.$$

If $\bar{x} \in \widetilde{M}$, then $\overline{\Delta b} = (A\bar{x} - b)^+$ will be the optimal increment of the vector b provided compatibility of the system $Ax \leq b + \overline{\Delta b}$ and, along with it, compatibility of (44.1) with the change b on $b + \overline{\Delta b}$.

44.3. Dual regularization of LP problems

Problems (44.4) constructed in the process of regularizing the initial problem L without assumption on its solvability arise indeed in the framework of duality theory for improper LP problems. The example of such problems is the following pair:

$$\max_{x \geq 0} \left\{ (c, x) - \varepsilon \|x\|^2 + \frac{1}{4\delta} \|(Ax - b)^+\|^2 \right\}, \tag{44.9}$$

$$\min_{u \geq 0} \left\{ (b, u) + \delta \|u\|^2 + \frac{1}{4\varepsilon} \|(c - A^T u)^+\|^2 \right\}, \tag{44.9#}$$

which is solvable always (independently of solvability of the problem L : $\max\{(c, x) \mid Ax \leq b, \ x \geq 0\}$ and L^* : $\min\{(b, u) \mid A^T u \geq c, \ u \geq 0\}$); and their optimal values coincide opt $L = $ opt L^*. This is a particular case of more general (relatively Theorem 25.2) duality theorem for improper problems from the book (Eremin, 1988) (see relations (25.12), (25.13) and Theorem 23.1 in that book). The last theorem belongs to A. A. Vatolin.

Problems (44.9) and (44.9#) we rewrite as follows:

$$- \min_{x \geq 0} \left\{ \varepsilon \left\| x - \frac{c}{2\varepsilon} \right\|^2 - \frac{1}{4\delta} \|(Ax - b)^+\|^2 \right\} + \frac{\|c\|^2}{4\varepsilon}, \tag{44.10}$$

$$\min_{u \geq 0} \left\{ \delta \left\| u + \frac{b}{2\delta} \right\|^2 + \frac{1}{4\varepsilon} \|(c - A^T u)^+\|^2 \right\} - \frac{\|b\|^2}{4\delta}, \tag{44.10#}$$

where opt (44.10) = opt (44.10#). Denoting the problems

$$\min_{x \geq 0} \left\{ \varepsilon \left\| x - \frac{c}{2\varepsilon} \right\|^2 - \frac{1}{4\delta\varepsilon} \|(Ax - b)^+\|^2 \right\}, \tag{44.11}$$

$$\min_{u \geq 0} \left\{ \delta \left\| u + \frac{b}{2\delta} \right\|^2 + \frac{1}{4\delta\varepsilon} \|(c - A^T u)^+\|^2 \right\}, \qquad (44.12)$$

we obtain the relations

$$\varepsilon \operatorname{opt}(44.11) + \delta \operatorname{opt}(44.12) = \frac{\|c\|^2}{4\varepsilon} + \frac{\|b\|^2}{4\delta},$$

$$\operatorname{Arg}(44.9) = \operatorname{Arg}(44.11), \qquad \operatorname{Arg}(44.9^\#) = \operatorname{Arg}(44.12).$$

As for solving problems (44.11), (44.12), we can use the methods developed in Section 43.

Chapter 7.

Piecewise linear functions and problems of disjunctive programming

---

45. INTRODUCTORY CONSIDERATIONS

The problems we investigate for piecewise linear programming naturally lead to setting the problem for disjunctive programming. An arbitrary continuous piecewise linear function (k-linear function or, simply, k-function) given in $\mathbb{R}^n$ admits the standard representation (Plotnikov, 1983)

$$f(x) := \min_{(j)} |A_j x - b^j|_{\max}, \qquad (45.1)$$

where A_j is a matrix; $x \in \mathbb{R}^n$, $b^j \in \mathbb{R}^{m_j}$; the symbol $|\cdot|_{\max}$ denotes the maximal coordinate of the vector standing inside $|\cdot|_{\max}$. The inequality $f(x) \leq 0$ defines the set $M = \bigcup_{(j)} M_j$, $M_j = \{x \mid A_j x \leq b^j\}$. An arbitrary finite system of inequalities consisted from k-functions can be reduced constructively to one inequality with the function $f(x)$ of (45.1) form. An arbitrary problem of piecewise linear programming admits the standard representation

$$\max\{(c, x) \mid f(x) \leq 0\}, \qquad (45.2)$$

where $f(x)$ is of (45.1) form. Unlike the traditional representation of admissible domain as intersection of a finite number of sets (half-spaces, simple convex sets and so on) in this case the admissible domain of optimization problem (45.2) is given by *conjunction* of the sets (polyhedrons), i.e.,

$$M = \{x \mid f(x) \le 0\} = \bigcup_{(j)} M_j, \qquad M_j = \{x \mid A_j x \le b^j\}.$$

Now we consider the general setting. Suppose $\{M_j\}_1^m$, are arbitrary sets from $\mathbb{R}^n$; $f(x)$ is an arbitrary function given in $\mathbb{R}^n$. We consider the two problems

$$P_\cap : \quad \max\left\{f(x) \mid x \in \bigcap_{j=1}^m M_j\right\}, \tag{45.3}$$

$$P_\cup : \quad \max\left\{f(x) \mid x \in \bigcup_{j=1}^m M_j\right\}. \tag{45.4}$$

The first problem we shall call *the problem in conjuctive setting*; the second problem we shall call *the problem in disjunctive setting*. Form (45.3) is usual for problems of mathematical programming. The second form corresponds to piecewise linear problem (45.2). We suppose that in (45.4)

$$M_j = \{x \mid F_j(x) \le 0\}, \qquad F_j : \mathbb{R}^n \to \mathbb{R}^{m_j}, \qquad j = 1, \dots, m.$$

The Lagrange function for the problem $P_\cap$ is the function

$$\Phi_\cap(x, u) = f(x) - \sum_{j=1}^m (u_j, F_j(x)). \tag{45.5}$$

The function

$$\Phi_\cup(x, u) = f(x) - \min_{(j)} (u_j, F_j(x)) \tag{45.6}$$

we shall call the *disjunctive Lagrange function* for the problem $P_\cup$. We fix the scheme of correspondence of the Lagrange functions to the problems $P_\cap$ and $P_\cup$:

$$P_\cap \ \rightarrow \ \Phi_\cap(x, u) = f(x) - \sum_{j=1}^{m} (u_j, F_j(x)),$$

$$P_\cup \ \rightarrow \ \Phi_\cup(x, u) = f(x) - \min_{(j)}(u_j, F_j(x)).$$

The disjunctive Lagrange function

$$L_\cup(x, u) = (c, x) - \min_{(j)}(u_j, A_j x - b^j) \tag{45.7}$$

will correspond to the problem of piecewise linear programming (k-problem) in standard form (45.2).

We can connect many problems related to k-problems with this function. Some of these problems we shall consider below. The algebra of k-functions and k-problems admits extension of their settings, namely, to the limits of *σ-extension* of linear functional spaces. We talk about algebraic extension of a fixed functional space $\mathbb{F}_0$ to a minimal linear space $\mathbb{F}$ closed relatively the operation of *discrete maximum*. This means that if $\{f_j(x)\} \subset \mathbb{F}$ then $\max_{(j)} f_j(x) \in \mathbb{F}$. If $\mathbb{F}_0$ is the class of linear functions, then $\mathbb{F}$ is the class of k-functions. If $\mathbb{F}_0$ is the class of square functions, then $\mathbb{F}$ is the class of piecewise square functions and so on. This property allows us to carry out some problems of piecewise programming and the disjunctive problems connected with them beyond the scope of linear settings.

Piecewise linear functions and their apparatus are of great importance, however, there are few works devoted to this problem. We mention the following (Volokitin, 1979), (Plotnikov, 1983), (Benchekroun and Falk, 1991), (Gorokhovik and Zor'ko, 1994), (Kripfganz and Schulze, 1987), and (Melzer, 1986).

46. σ-EXTENSIONS OF LINEAR FUNCTIONAL SPACES

We suppose that $\mathbb{F}_0$ is a certain linear functional space with the real space $\mathbb{X}$ of values of argument of functions from $\mathbb{F}_0$. If $\{f_j(x)\}_{j \in J} \subset \mathbb{F}_0$, then the function of discrete maximum $f(x) := \max_{j \in J} f_j(x)$ may belong to $\mathbb{F}_0$ and not belong to $\mathbb{F}_0$. The way of forming the function $f(x)$ we shall call the σ-operation.

We shall consider the minimal extension of the space $\mathbb{F}_0$ to the linear space $\mathbb{F}$ provided the property of σ-completeness:

$$\{f_j(x)\}_{j \in J} \subset \mathbb{F} \implies \max_{j \in J} f_j(x) \in \mathbb{F}. \tag{46.1}$$

In this case, the property of linear completeness evidently holds

$$\{f_j^i \mid i \in I, \; j \in J_i\} \subset \mathbb{F} \implies \sum_{i \in I} \alpha_i \max_{j \in J_i} f_j^i \in \mathbb{F} \tag{46.2}$$

where $\alpha_i \in \mathbb{R}$, $i \in I$.

The minimal σ-complete extension we shall call σ-extension. From the sense of such extension we see that

$$\mathbb{F} = \bigcup_{k=0}^{+\infty} \mathbb{F}_k,$$

where

$$\mathbb{F}_{k+1} = \left\{ \sum_{i \in I} \alpha_i \max_{j \in J_i} f_j^i \;\middle|\; \begin{array}{l} f_j^i \in \mathbb{F}_k, \; \alpha_i \in \mathbb{R}, \; i \in I, \; j \in J_i, \\ |I| < +\infty, \; |J_i| < +\infty \end{array} \right\}.$$

Indeed, all the functions from $\mathbb{F}$ can be transformed to a certain standard form. In the basis of such transformation we have certain identities valid for an arbitrary collection of functions

$$\max_{j \in J} f_j + \max_{i \in I} g_i = \max_{(j,i) \in J \times I} (f_j + g_i); \tag{46.3}$$

$$\sum_{i \in I} \alpha_i \max_{j \in J_i} f_j^i = \max_{s_i \in J_i} \sum_{i \in I_+} \alpha_i f_{s_i}^i - \max_{s_i \in J_i} \sum_{i \in I_-} |\alpha_i| f_{s_i}^i. \tag{46.4}$$

Here $I_+ = \{i \mid \alpha_i > 0\}$, $I_- = \{i \mid \alpha_i < 0\}$,

$$\max_{j=1,\dots,m} f_j = \left[\max_{j=1,\dots,m-1} f_j - f_m \right]^+ + f_m$$
$$= \left[f_m - \max_{j=1,\dots,m-1} f_j \right]^+ + \max_{j=1,\dots,m-1} f_j; \tag{46.5}$$

$$\min_{i=1,\dots,n} f_i = -\left[\min_{j=1,\dots,n-1} f_i - f_n \right]^+ + \min_{i=1,\dots,n-1} f_i$$
$$= -\left[f_n - \min_{i=1,\dots,n-1} f_i \right]^+ + f_n; \tag{46.6}$$

$$\left[\max_{j \in J} f_j - \max_{i \in I} g_i\right]^+ = \max_{(j,i) \in J \times I}\{f_j, g_i\} - \max_{i \in I} g_i; \tag{46.7}$$

$$\max_{j \in J} f_j - \max_{i \in I} g_i = \min_{i \in I} \max_{j \in J}(f_j - g_i) = \max_{j \in J} \min_{i \in I}(f_j - g_i). \tag{46.8}$$

All these identities are general (they holds for arbitrary functions) and can be verified immediately. Parallel with $\mathbb{F}$ we introduce the space

$$\mathbb{H} = \bigcup_{i=0}^{+\infty} \mathbb{H}_i,$$

where $\mathbb{H}_0 = \mathbb{F}_0$, $\mathbb{H}_{k+1} = \left\{ \sum_{i \in I} \alpha_i f_i^+ \;\middle|\; \alpha_i \in \mathbb{R}, \; f_i \in \mathbb{H}_k, \; |I| < +\infty \right\}$.

The operation of positive cut "+" is a particular case of σ-operation, however, multiple applied, it gives the same class $\mathbb{F}$. The following theorem holds.

Theorem 46.1. *1) All functions from $\mathbb{F}$ can be reduced to each of the standard forms*

$$\max_{j \in J} f_j - \max_{i \in I} g_i, \tag{46.9}$$

$$\min_{i \in I} \max_{j \in J_i} f_j^i, \tag{46.10}$$

$$\max_{j \in J} \min_{i \in I_j} f_i^j, \tag{46.11}$$

where $\{f_j, g_i, f_j^i\} \subset \mathbb{F}_0$; (46.9) yields $\mathbb{F}_k = \mathbb{F}_1$ for $k > 1$; therefore, $\mathbb{F} = \mathbb{F}_1$.
2) The classes $\mathbb{F}$ and $\mathbb{H}$ coincide.

Proof. 1) Relation (46.9) denotes coincidence $\mathbb{F}_k$ with $\mathbb{F}_1$ for $k > 1$. Indeed, it suffices to show that $\mathbb{F}_2 = \mathbb{F}_1$. By (46.4), we can restrict ourselves by the transformation of the function $f(x) = \max_{i \in I} f_i$ for $\{f_i\}_I \subset \mathbb{F}_1$ to the form (46.9). Let $I = \{1, \ldots, n\}$. We shall prove by the induction method with respect to n. If $n = 1$, then $f(x) = f_1(x)$ satisfies the required property of representation (46.9). Suppose now that $n > 1$. As $f_i \in \mathbb{F}_1$, then these functions can be represented in the form

$$f_i = \max_{j \in J_i} f_j^i - \max_{k \in I_i} g_k^i$$

for $\{f_j^i, g_k^i\} \subset \mathbb{F}_0$. We have

$$f(x) \overset{(46.5)}{=} \left[\max_{i=1,\dots,n-1} f_i - f_n \right]^+ + f_n.$$

By the induction method, we have

$$\max_{i=1,\dots,n-1} f_i =: \bar{f} \in \mathbb{F}_1.$$

As $f = [\bar{f} - f_n]^+ + f_n$, where $\{\bar{f}, f_n\} \subset \mathbb{F}_1$, then, by (46.4), we have $\bar{f} - f_n \in \mathbb{F}_1$. Taking into account (46.7), we obtain $[\bar{f} - f_n]^+ \in \mathbb{F}_1$, which $f_n \in \mathbb{F}_1$ gives $f \in \mathbb{F}_1$. So, we have proved that $\mathbb{F}_2 = \mathbb{F}_1$; therefore, $\mathbb{F} = \mathbb{F}_1$.

Representation of the function from $\mathbb{F}$ in form (46.10) or (46.11) follows from identity (46.8).

2) We shall establish first inclusion $\mathbb{H} \subset \mathbb{F}_1$, i.e., $\mathbb{H}_k \subset \mathbb{F}_1$, $\forall k$. If $f \in \mathbb{H}_1$, then, by (46.4), we have $f \in \mathbb{F}_1$. Therefore, $\mathbb{H}_1 \subset \mathbb{F}_1$. Suppose now that $\mathbb{H}_k \subset \mathbb{F}_1$. We shall prove that $\mathbb{H}_{k+1} \subset \mathbb{F}_1$. An arbitrary function from $\mathbb{H}_{k+1}$ has the form

$$f = \sum_{i \in I} \alpha_i f_i^+, \qquad \{f_i\} \subset \mathbb{H}_k \subset \mathbb{F}_1.$$

Hence it follows that f is represented as a linear combination of functions with discrete maximum with generators from $\mathbb{F}_0$. Taking into account (46.4), this gives us the required representation (46.9) for f.

Conversely, we suppose that $f \in \mathbb{F}$, i.e., f is of the form (46.9). If we show that the function of discrete maximum with generators belonging to $\mathbb{F}_0$ belongs to $\mathbb{H}$, i.e., to a certain $\mathbb{H}_k$, we shall show that $f \in \mathbb{H}$. So, we suppose that

$$f = \max_{j \in J} f_j, \qquad \{f_j\} \subset \mathbb{F}_0, \qquad j = 1, \dots, m.$$

If $m = 1$, then $f = f_1 \subset \mathbb{H}_0$. Let $m > 1$. We write relation (46.5)

$$f = \left[\max_{j=1,\dots,m-1} f_j - f_m \right]^+ + f_m.$$

If, by the induction supposition $\bar{f} = \max_{j=1,\dots,m-1} f_j \in \mathbb{H}_k$, then $f = [\bar{f} - f_m]^+ + f_m \in \mathbb{H}_{k+1}$ which was to be proved.

Representations (46.9)–(46.11) of functions from the class $\mathbb{F}$ are, thus, equivalent. Constructive proof of this fact is based on identities (46.3)–(46.8).

We have noted that representation of a function in form (46.9) can be rewritten in form (46.10) for $f_j^i = f_j - g_i$ (see, (46.8)). We must ensure in inverse.

So, we suppose that f is of form (46.10), i.e., $f = \min_{i \in I} \max_{j \in J_i} f_j^i$. If $I = \{1, \ldots, n\}$ and $n = 1$, then $f = \max_{j \in J_1} f_j^1 \in \mathbb{F}_1$. For $n > 1$ we can use the induction method as above. By (46.6), we have

$$f = \Big[\min_{i=1,\ldots,n-1} \max_{j \in J_i} f_j^i - \max_{j \in J_n} f_j^n \Big]^+ + \max_{j \in J_n} f_j^n.$$

By the inductive supposition,

$$\bar{f} = \min_{i=1,\ldots,n-1} \max_{j \in J_i} f_j^i \in \mathbb{F}_1,$$

i.e., $\bar{f}$ can be represented in form (46.9). Then, using transformations (46.3), (46.4), and (46.7), we transform the function f to form (46.9) which was to be proved.

Equivalence of representations (46.11) and (46.9) is established analogously. The theorem is proved. $\square$

The functions f from $\mathbb{F}$ we shall call by *σ-functions* or *σ-piecewise functions*. If $\mathbb{F}_0$ is a space of linear functions, then $\mathbb{F}$ is a space of piecewise linear functions.

47. THE PROBLEM ON THE SADDLE POINT OF THE DISJUNCTIVE LAGRANGE FUNCTION

We consider here problems (45.3) and (45.4) with their Lagrange functions (45.5) and (45.6) respectively. Problem (45.4), where

$$M_j = \{ x \mid F_j(x) \le 0 \}, \qquad j = 1, \ldots, m$$

can be rewritten in the form

$$\max \{ f(x) \mid \min_{(j)} |F_j(x)|_{\max} \le 0, \ x \ge 0 \}. \tag{47.1}$$

Here the restrictions of problem (45.4) we supplemented by the condition of non-negativeness of the vector x, i.e., $x \ge 0$. This is connected with symmetry in setting the dual problems. The sets M_j will be the sets

$$\{ x \mid F_j(x) \le 0, \ x \ge 0 \}, \qquad j = 1, \ldots, m.$$

If $\Phi_{\cup}(x, u)$ is a function of form (45.6), then, with (47.1) we connect the saddle problem $[\bar{x}, \bar{u}] \geq 0$

$$\Phi_{\cup}(x, \bar{u}) \underset{\forall x \geq 0}{\leq} \Phi_{\cup}(\bar{x}, \bar{u}) \underset{\forall u \geq 0}{\leq} \Phi_{\cup}(\bar{x}, u). \tag{47.2}$$

In mathematical programming the following fact is well known and can be easily established (Arrow *et al.*, 1958):

if $[\bar{x}, \bar{u}] \geq 0$ is the saddle point of the Lagrange function $\Phi_{\cap}(x, u)$ correspondent to problem (45.3), then $(\bar{u}_j, F_j(\bar{x})) = 0$, $\forall j$ and $\bar{x} \in$ Arg (43.5).

For problem (47.1) and its disjunctive Lagrange function $\Phi_{\cup}(x, u)$ the analogous statement holds.

Theorem 47.1. *If $[\bar{x}, \bar{u}] \geq 0$ is the saddle point for $\Phi_{\cup}(x, u)$, then $\min_{(j)}(\bar{u}_j, F_j(\bar{x})) = 0$ and $\bar{x} \in$ Arg (47.1).*

Proof. Following the definition of the saddle point, we have

$$\Phi_{\cup}(x, \bar{u}) \underset{\forall x \geq 0}{\leq} \Phi_{\cup}(\bar{x}, \bar{u}) \underset{\forall u \geq 0}{\leq} \Phi_{\cup}(\bar{x}, u),$$

or

$$f(x) - \min_{(j)}(\bar{u}_j, F_j(x)) \underset{\forall x \geq 0}{\leq} f(\bar{x}) - \min_{(j)}(\bar{u}_j, F_j(\bar{x})), \tag{47.3}$$

$$f(\bar{x}) - \min_{(j)}(\bar{u}_j, F_j(\bar{x})) \underset{\forall u \geq 0}{\leq} f(\bar{x}) - \min_{(j)}(u_j, F_j(\bar{x})). \tag{47.4}$$

Relation (47.4) we rewrite in the form

$$\min_{(j)}(\bar{u}_j, F_j(\bar{x})) \underset{\forall u \geq 0}{\leq} \min_{(j)}(u_j, F_j(\bar{x})). \tag{47.5}$$

First, we prove that $\bar{x} \in M = \bigcup_{(j)} M_j$, i.e., $\exists j_0 : F_{j_0}(\bar{x}) \leq 0$. If $F_j(\bar{x}) \not\leq 0$, $\forall j$, then, choosing $u \geq 0$ we can make the right-hand side value in (47.5) arbitrarily large, which would contradict relation (47.5). This gives $\alpha := \min_{(j)}(\bar{u}_j, F_j(\bar{x})) \leq 0$. Indeed, $\alpha = 0$. If $\alpha < 0$, then relation (47.5) for $u = 0$ would generate the contradictory inequality $0 > -|\alpha| \geq 0$. So, the relation $\min_{(j)}(\bar{u}_j, F_j(\bar{x})) = 0$ is established.

Now, it is necessary to show that $\bar{x}$ is optimal for (47.1), i.e., $f(x) \leq f(\bar{x})$, $\forall x \in M$. Consider relation (47.3), which we rewrite in the form

$$f(\bar{x}) - f(x) \geq - \min_{(j)}(\bar{u}_j, F_j(x)). \tag{47.6}$$

If $x \in M$, i.e., $x \in M_{j_0}$ for a certain j_0, then $\min_{(j)}(\bar{u}_j, F_j(x)) \leq 0$. Taking into account (47.6), this gives $f(\bar{x}) - f(x) \geq 0$, $\forall x \in \bigcup_{(j)} M_j$. The theorem is proved. $\square$

In mathematical programming (MP) the most important is the inverse theorem, formulated the conditions which guarantee the existence of saddle point for corresponding Lagrange function. Such theorems hold for solvable LP problems, convex programming (CP) with conditions of regularity and for certain classes of MP problems. For disjunctive settings of MP problems the situations is more complicated. In particular, we strengthen the conditions of regularity.

Problem (47.1) we shall call completely regular if each of problems

$$\max\{f(x) \mid F_j(x) \leq 0, \ x \geq 0\}, \quad j = 1, \ldots, m \tag{47.7}$$

is solvable. The equivalent condition of such regularity is solvability of problem (45.4) and non-emptiness of all M_j : $M_j \neq \varnothing$, $j = 1, \ldots, m$.

Suppose now that

$$\Phi_j(x, u_j) = f(x) - (u_j, F_j(x))$$

is the Lagrange function for (47.7). Here $x \in \mathbb{R}^n$, $u_j \in \mathbb{R}^{m_j}$.

Theorem 47.2. Let each of the functions $\Phi_j(x, u_j)$ have the saddle point $[\bar{x}_j, \bar{u}_j] \geq 0$, $j = 1, \ldots, m$. If $\bar{x}$ is those value $\bar{x}_j$ where $\max_{(j)} f(\bar{x}_j)$ achieves, then

$$\Phi_\cup(x, \bar{u}) \underset{\forall x \geq 0}{\leq} f(\bar{x}), \tag{47.8}$$

where

$$\Phi_\cup(x, \bar{u}) = f(x) - \min_{(j)}(\bar{u}_j, F_j(x)).$$

Proof. First, we have $(\bar{u}_j, F_j(\bar{x})) = 0$, $j = 1, \ldots, m$ and

$$f(x) - (\bar{u}_j, F_j(x)) \leq f(\bar{x}_j) \qquad (\leq f(\bar{x})), \qquad \forall x \geq 0.$$

Hence we have $\max_{(j)}[f(x) - (\bar{u}_j, F_j(x))] \leq f(\bar{x})$. As the left-hand side of this inequality is equal to $f(x) - \min_{(j)}(\bar{u}_j, F_j(x))$ $(= \Phi(x, \bar{u}))$ then inequality (47.8) holds. The theorem is proved. $\square$

Now, we modify the function $\Phi_\cup(x, u)$. Namely, we set

$$\overline{\Phi}_\cup(x, u) = f(x) - \min_{(j)}(u_j, F_j^+(x)),$$

where "+" denotes the positive cut. The following theorem holds.

Theorem 47.3.

1) For $\overline{\Phi}_\cup(x, u)$ Theorems 47.1 and 47.2 hold.

2) In conditions of Theorem 47.2 the point $[\bar{x}, \bar{u}]$ *(from this theorem)* realizes the saddle point for $\overline{\Phi}_\cup(x, u)$.

Proof. As for statements from 1), they are established similarly as in Theorems 47.1 and 47.2. We consider the statement 2). We need to show that

$$f(x) - \min_{(j)}(\bar{u}_j, F_j^+(x)) \underset{\forall x \geq 0}{\leq} f(\bar{x}) - \min_{(j)}(\bar{u}_j, F_j^+(\bar{x}))$$

$$\underset{\forall u \geq 0}{\leq} f(\bar{x}) - \min_{(j)}(u_j, F_j^+(\bar{x})). \qquad (47.9)$$

As $\bar{x}$ realizes $\max_{(j)} f_j(\bar{x}_j)$, let $\bar{x} = \bar{x}_{j'}$. Then $F_{j'}^+(\bar{x}) = 0$; therefore, $\min_{(j)}(u_j, F_j(\bar{x})) = 0$ for each $u_j \geq 0$, $\forall j$. This means, in particular, that $\min_{(j)}(\bar{u}_j, F_j^+(\bar{x})) = 0$. Therefore, the right inequality in (47.9) holds evidently for each $u_j \geq 0$ and the left inequality repeates the inequality (47.8) proved for the case of the function $\overline{\Phi}_\cup(x, u)$. The theorem is proved. $\square$

Remark to the formulations of Theorems 47.1–47.3. These theorems are of general character. In these theorems the setting of problems (47.7) is not revises. But if, for example, (47.7) is the problem of convex programming, then, instead of postulating the existence of saddle point for their Lagrange functions (in Theorem 47.2) we can suggest solvability of problems (47.7) and validity of the condition $\exists p \geq 0$, $F_j(p) < 0$ (condition of regularity). This condition would guarantee existence of the saddle point $\Phi_j(x, u)$, $j = 1, \ldots, m$. As for the case of linear setting of problem (47.1), we shall revise below certain theorems. Note that the linear case has its own interest since the problems of piecewise linear programming are actual.

Let

$$\max\{f(x) \mid f_j(x), \ j = 1, \ldots, m, \ x \geq 0\}$$

be the general setting for a piecewise linear programming problem. This means that $\{f, f_j\}_j$ are piecewise linear functions. Rewrite it in the equivalent form

$$\max\{t \mid t \leq f(x), \ f_j(x) \leq 0, \ j = 1, \ldots, m, \ x \geq 0\},$$

and in the form

$$\max\left\{t \mid \varphi(x, t) := (t - f(x))^+ + \sum_{j=1}^{m} f_j^+(x) \leq 0, \ x \geq 0\right\}.$$

As $\varphi(x, t)$ is piecewise linear function also dependent on $z = [x, t]$, this problem can be reduced to form (46.10) correspondent to the case when $\mathbb{F}_0$ is the space of linear functions. So, we have noted that an arbitrary problem of piecewise linear programming (k-problem) can be written in the form

$$L_\cup : \ \max\{(c, x) \mid \min_{(j)} |A_j x - b^j|_{\max} \leq 0, \ x \geq 0\}, \tag{47.10}$$

or

$$\max\left\{(c, x) \mid x \in \bigcup_{j=1}^{m} M_j\right\},$$

where $M_j = \{x \geq 0 \mid A_j(x) \leq b_j\}$, $j = 1, \ldots, m$.

Further, we shall consider the problem (47.10) and its Lagrange function

$$L_\cup(x, u) := (c, x) - \min_{(j)}(u_j, A_j x - b_j).$$

Lemma 47.1. *Let problem (47.10) be solvable and $M_j \neq \varnothing$, $\forall j$ (i.e. is completely regular). Then the function $L_\cup(x, u) = (c, x) - \min_{(j)}(u_j, A_j x - b_j)$ has the saddle point $[\bar{x}, \bar{u}]$, where*

$$\bar{x} = \arg\max_{(j)}(c, \bar{x}_j), \qquad \bar{x}_j \in \text{Arg}\max_{x \in M_j}(c, x),$$

$$\bar{u}_j \in \text{Arg}\min_{u \in M_j^*}(b^j, u), \qquad j = 1, \dots, m,$$

where $M_j^* := \{u \geq 0 \mid A_j^T u \geq c\}$.

Proof. If we show that $\alpha := \min_{(j)}(\bar{u}_j, A_j\bar{x} - b_j) = 0$, then, by Theorem 47.2, the left inequality in the definition of saddle point for $L_\cup(x, u)$ will hold true. As the vector $\bar{x} \geq 0$ satisfies one of the systems $A_j x \leq b^j$, then $\alpha \leq 0$. We show that $\alpha \geq 0$. We can write

$$\min_{(j)}(\bar{u}_j, A_j\bar{x} - b^j) = \min_{(j)}\left[-(b^j, \bar{u}_j) + (c, \bar{x}) + (A_j^T\bar{u}_j - c, \bar{x})\right]$$

$$\geq \min_{(j)}[(c, \bar{x}) - (c, \bar{x}_j)] \geq 0.$$

We have used above the relations: $A_j^T\bar{u}_j - c \geq 0$; $(b^j, \bar{u}_j) = (c, \bar{x}_j)$ by the duality theorem in LP; $(c, \bar{x}) \geq (c, \bar{x}_j)$, $\forall j$.

It remains to establish the right inequality in the definition of saddle point for $L_\cup(x, u)$, i.e.,

$$L_\cup(\bar{x}, \bar{u}) \underset{\forall u \geq 0}{\leq} L_\cup(\bar{x}, u).$$

Taking into account $\alpha = 0$, the inequality becomes as follows:

$$0 \underset{\forall u_j \geq 0}{\leq} -\min_{(j)}(u_j, A_j\bar{x} - b^j). \qquad (47.11)$$

As $\exists j_0 : A_{j_0}\bar{x} - b^j \leq 0$, then for each $u \geq 0$:

$$\min_{(j)}(u_j, A_j\bar{x} - b^j) \leq 0;$$

therefore, (47.11) holds. The lemma is proved. $\quad\square$

This lemma and Theorem 47.1 yield the following theorem.

Theorem 47.4. *If the systems* $A_j x \leq b^j$, $x \geq 0$, $j = 1, \ldots, m$ *are compatible then problem* (47.10) *is solvable if and only if its disjunctive Lagrange function*

$$L_\cup(x, u) := (c, x) - \min_{(j)}(u_j, A_j x - b_j)$$

has the saddle point $[\bar{x}, \bar{u}] \geq 0$. *In this case*

1) *if* $L_\cup$ *is solvable and* $\bar{x}_j \in \text{Arg} L_j$, $\bar{u}_j \in \text{Arg} L_j^*$, $\bar{x} = \arg\max_{(j)}(c, \bar{x}_j)$, *then* $[\bar{x}, \bar{u}]$ *is the saddle point;*

2) *if* $[\bar{x}, \bar{u}]$ *is the saddle point for* $L_\cup(x, u)$, *then* $\{\bar{x}, \bar{u}_j\}$ *satisfies all the relations from 1).*

48. PIECEWISE LINEAR FUNCTIONS AND SYSTEMS OF PIECEWISE LINEAR INEQUALITIES

Piecewise linear functions are σ-functions in the case when $\mathbb{F}_0$ is the space of linear functions. We can define piecewise linear functions in two ways: either as it was done in Section 46, or using a certain exiomatics identifying such functions. We consider the second approach.

Let us have a finite collection of polyhedrons $\{M_j\}_J$ and eigen linear functions $\{l_j(x)\}_J$. We shall say that the system $\{M_j, l_j(x)\}$ sets an one-to-one piecewise linear function $l(x)$ given in $\mathbb{X}$ if

1) $\bigcup_{j \in J} M_j = \mathbb{X}$, $M_i^0 \cap M_j^0 = \varnothing$ for $i \neq j$;

2) $l(x) \equiv l_j(x)$, $\forall x \in M_j$, $\forall j \in J$.

Here M_j^0 is the algebraic interioe of the polyhedron M_j, i.e., $y \in M_j^0 \iff y + ts \in M_j$ for each $s \in \mathbb{X}$ and $t \geq 0$ sufficiently small. *Polyhedron*, as before, denotes the set given by a finite system of proper linear inequalities

$$(f_j, x) - \alpha_j \leq 0, \qquad j \in J.$$

In this definition some M_j or M_j^0 may be empty.

We shall use the notations: $\mathbb{L}_0$ is the space of linear functions, $\mathbb{L}$ is the space of k-linear functions defined in external way following the property 1) and 2). Indeed, the class of functions $\mathbb{L}$ coincides with the class $\mathbb{F}$ from the previous paragraph which is constructed from $\mathbb{F}_0 = \mathbb{L}$. This means that

$\mathbb{L}$ is the minimal extension of the space of linear functions closed relatively operations of discrete maximum (see Section 46). Therefore, representation (46.9) and each of (46.10) and (46.11) is universal representation of piecewise linear functions (we shall say briefly, k-functions).

For unification and simplification of record of k-functions, the systems of inequalities consisted from k-functions, the problems of piecewise linear programming and so on, we shall set further $\mathbb{X} = \mathbb{R}^n$. Then

$$\mathbb{L}_0 = \{l(x) = (a, x) - \alpha \mid a \in \mathbb{R}^n, \ \alpha \in \mathbb{R}\},$$
$$\mathbb{L} = \{\max_{j \in J} l_j(x) - \max_{i \in I} h_i(x) \mid \{l_j, h_i\} \subset \mathbb{L}_0, \ |J| < +\infty, \ |I| < +\infty\}.$$

Let

$$|z|_{\max} = \max_{(i)} z_i, \qquad |z|_{\min} = \min_{(i)} z_i,$$

where z is the vector from a finite-dimensional space. If $Ax - b = [l_1(x), \ldots, l_m(x)]^T$ is the vector of linear functions, then, following the introduced notations, we shall have

$$|Ax - b|_{\max} = \max_{(i)} l_i(x).$$

Representations of piecewise linear functions in form (46.9)–(46.11) attain the universal form

$$|Ax - b|_{\max} - |Bx - d|_{\max}, \tag{48.1}$$
$$\min_{(i)} |A_i x - b^i|_{\max}, \tag{48.2}$$
$$\max_{(j)} |A_j x - b^j|_{\min}. \tag{48.3}$$

Note also the properties of the functions of discrete maximum

$$|z|_{\max} = -|-z|_{\min};$$
$$\text{for } \alpha > 0: \quad |\alpha z|_{\max} = \alpha |z|_{\max};$$
$$\text{for } \alpha < 0: \quad |\alpha z|_{\max} = \alpha |z|_{\min}.$$

A finite system of inequalities consisted of piecewise linear functions (k-functions) can be written as follows:

$$\min_{(j)} |A_j^t x - b_t^j|_{\max}, \qquad t = 1, \ldots, T. \tag{48.4}$$

It can be rewritten by means of the only inequality

$$\sum_{t=1}^{T} \left(\min_{(j)} |A_j^t x - b_t^j|_{\max} \right)^+ \leq 0,$$

or, (on the basis of Theorem 46.1), in the form

$$\min_{j=1,\ldots,m} |A_j x - b^j|_{\max} \leq 0. \tag{48.5}$$

This setting we shall consider as one the standard setting. Another standard setting is

$$|Ax - b|_{\max} - |Bx - d|_{\max} \leq 0. \tag{48.6}$$

So, an arbitrary finite system of piecewise linear inequalities can be transformed to each of standard forms (48.4)–(48.6).

We consider now the representation of the system in form (48.5). We set $M_j = \{x \mid A_j x \leq b^j\}$. Then the set of solutions of inequality (48.5) will be $M = \bigcup_{j=1}^{m} M_j$. On the other hand, if M is an arbitrary polyhedron set, suppose from $\mathbb{R}^n$, i.e., $M = \bigcup_{j=1}^{m} M_j$ and $\{M_j\}$ are the polyhedrons

$$M_j = \{x \mid A_j x \leq b^j\},$$

then M is the set of solutions of inequality (48.5).

Consider inequality (48.6) from the similar point of view. Let

$$Ax - b = [l_1(x), \ldots, l_m(x)]^T, \qquad Bx - d = [s_1(x), \ldots, s_k(x)]^T,$$
$$\{l_j(x), s_i(x)\}_{1,1}^{m,k} \subset \mathbb{L}_0$$

($\mathbb{L}_0$ is the space of linear functions). We set

$$M_i = \{x \mid l_j(x) \leq s_i(x), \ j = 1, \ldots, m\}.$$

Then, as is easy to see, $\bigcup_{i=1}^{s} M_i$ coincides with the set of solutions of inequality (48.6). Thus, for inequality (48.6), we have selected those polyhedrons M_i whose union gives the set of solutions of inequality (48.6). On the other hand, if a polyhedral set M is given, for example, as in previous case – by the collection of systems of linear inequalities each of which sets a convex polyhedral component of the set M, then we shall need a chain of transformations of type (46.3)–(46.8) which leads to setting M by one inequality of type (48.6).

Finally, we consider the system of piecewise linear inequalities in form (48.4), formally more complicated then (48.5) or (48.6). We set

$$M_j^t := \{x \mid A_j^t x \le b_t^j\}, \qquad M_t := \bigcup_{(j)} M_j^t, \qquad M := \bigcap_{(t)} M_t.$$

The set M_t is the set of solutions of t inequality in system (48.4), so, M is the set of solutions of the whole system. The contents of Sections 46 and 48 show us the way of constructive relation between the polyhedral sets and their analytic setting. The algebra of transformations which accompany this relation may be sufficiently tedious, however, the logic of these transformations is simple and may be realized in computer in real situations.

49. THE PROBLEM OF PIECEWISE LINEAR PROGRAMMING

49.1. Preliminary remarks

An arbitrary problem of piecewise linear programming, i.e., the problem of determining the extremum of a k-function for restrictions given in the form of a finite system of inequalities with k-functions in left-hand sides can be written in a universal simple form

$$P: \quad \max\{(c, x) \mid \min_{j=1,\dots,m} |A_j x - b^j|_{\max} \le 0, \ x \ge 0\}. \tag{49.1}$$

If $f(x)$ is an arbitrary optimized (suppose maximized) k-function with one k-inequality, suppose $g(x) \le 0$ (possibly with $x \ge 0$), then rewriting the problem $\max\{f(x) \mid g(x) \le 0\}$ in the form

$$\max\{t \mid g(x) \le 0, \ f(x) \ge t\}$$

and transforming the system consisted of two k-inequalities to one k-inequality, we obtain the problem of search of maximum of a linear function for one restriction in the form of k-inequality.

So, the object of consideration we take problem (49.1). We introduce the partial problem (subproblem)

$$L_j : \max\{(c, x) \mid A_j x \leq b^j, \ x \geq 0\}. \tag{49.2}$$

The connection between (49.1) and L_j is as follows:

$$\text{opt}(49.1) = \max_{(j \,:\, M_j \neq \varnothing)} \text{opt}\, L_j, \tag{49.3}$$

where $M_j = \{x \geq 0 \mid A_j x \leq b^j\}$. In (49.2) for certain j the sets $M_j = \{x \geq 0 \mid A_j x \leq b^j\}$ may be empty in the case when initial problem (49.1) is solvable. This denotes that an arbitrary k-problem transformed to form (49.1) falles into a finite number of LP problems whose solutions give us the solution of the initial problem. As the correspondent transformations are constructive, the formal solution of an arbitrary problem of piecewise linear programming is reduced to using the construction and a certain method (for example, simplex–method) of solution of LP problem.

49.2. Solvability conditions for the k-problem

For k-problems some properties and theorems formulated in the framework of LP theory hold. Some of them are the corollaries of theorems from LP theory, some of them must be revised or proved. The following theorem needs no special proof.

Theorem 49.1 [on attainability]. *If*

$$\sup\{(c, x) \mid \min_{(j)} |A_j x - b^j|_{\max} \leq 0, \ x \geq 0\} < +\infty,$$

then the operation sup *in this problem is accessible.*

We write now the problem dual to L_j:

$$L_j^* : \min\{(b^j, u^j) \mid A_j^T u^j \geq c, \ u^j \geq 0big\}. \tag{49.4}$$

Suppose $M_j^* = \{u^j \geq 0 \mid A_j^T u^j \geq c\}$.

Theorem 49.2. *Problem* (49.1) *is solvable if and only if*

$$M = \bigcup_{j=1}^{m} M_j \neq \varnothing \quad \text{and} \quad M_j \neq \varnothing \implies M_j^* \neq \varnothing.$$

Proof. As the conditions $M_j \neq \varnothing$ and $M_j^* \neq \varnothing$ are necessary and sufficient for solvability of the problem L_j, then $\max\limits_{j \,:\, M_j \neq \varnothing} \text{opt}\, L_j$ is finite and is the value opt P. Conversely, if the problem P is solvable, then all the problems L_j with $M_j \neq \varnothing$ are solvable (i.e. the case opt $L_j = +\infty$ is impossible). Then, following the duality theorem in LP, we have $M_j^* \neq \varnothing$ which was to be proved. $\square$

Remark 49.1. Another variant of formulation of Theorem 49.2 is based on the equivalence

$$P \text{ is solvable} \quad \Longleftrightarrow \quad (M \neq \varnothing \ \& \ (M_j \neq \varnothing \implies L_j \text{ is solvable})).$$

49.3. Duality

Initial k-problem we take in form (49.1). We only add one supposition (not essential): dimensions of the vectors $A_j x - b^j$ are equal, i.e. the number of inequalities in each system $A_j x \leq b^j$ is the same. This condition allows us to denote the dual variable u^j for the problem (49.4) by u. In this case, we rewrite problem (49.4) in the form

$$\min\{(b^j, u) \mid A_j^T u \geq c, \ u \geq 0\}, \qquad j = 1, \ldots, m. \tag{49.5}$$

The following problem

$$P^* : \quad \max_{j \,:\, M_j^* \neq \varnothing} \min\{(b^j, u) \mid A_j^T u \geq c, \ u \geq 0\} \tag{49.6}$$

we shall consider as one of variants of the problem dual to P.

Relatively problem (49.6) the following theorem holds.

Theorem 49.3. *Problem (49.6) is solvable if and only if*

$$\exists j \in \overline{1, m} : \quad M_j^* \neq \varnothing \quad \& \quad M_j \neq \varnothing. \tag{49.7}$$

Proof. Really, if $J := \{j \mid M_j \neq \varnothing, M_j^* \neq \varnothing\}$ then for $j \notin J$: $\inf\{(b^j, u) \mid x \in M_j^*\} = -\infty$; therefore, we may take maximum in (49.6) only with respect to $j \in J$. This provides solvability of the problem P^*. Necessity of conditions (49.7) is evident: if P^* is solvable, then, at least for one j' the problem L_j^* is solvable, which is equivalent to (49.7). $\quad\square$

Theorems 49.2 and 49.3, and the duality theorem for LP problems yield

Theorem 49.4. *If problem (49.1) is solvable then (49.6) is solvable also, and their optimal values coincide.*

In the case of problems P and P^* the property of their simultaneous solvability or unsolvability, unlike LP, does not hold. Taking into account the possibility of improper optimal values we rewrite these problems in the form

$$
\begin{aligned}
P : \quad &\sup_{(j)} \inf_{x \in M_j} (c, x), \\
P^* : \quad &\sup_{(j)} \inf_{u \in M_j^*} (b^j, u).
\end{aligned}
$$

Conditions of simultaneous solvability of P and P^* are determined by Theorems 49.2 and 49.3 but the situation when P is unsolvable and P^* is solvable is possible. This corresponds to the case that if P^* is solvable; there $\exists j_0 : M_{j_0} \neq \varnothing$ & $M_{j_0}^* = \varnothing$, i.e. L_{j_0} is an improper problem of the second kind. As in this case $\sup_{x \in M_{j_0}}(c, x) = +\infty$, then $\operatorname{opt} P = +\infty$, i.e. P is unsolvable. Simultaneous unsolvability of problems P and P^* is realized on the pair of LP problems L and L^* which are improper of the third kind.

Construction of the dual problem P^* in form (49.6) is prompted by considerations coming from the dual transfer $L_j \xrightarrow{(*)} L_j^*$ and necessity of the general property of duality in MP, namely: $\operatorname{opt} P = \operatorname{opt} P^*$. To provide the last relation in P^* we have introduced the operation $\max_{j : M_j^* \neq \varnothing} \cdots$, however, the symmetry concept requires to have the general operation min in the dual problem. In this sense the construction of the problem P^* looks rather artificial. Here another approach is possible. This approach is connected with known general situation relatively duality, when the initial problem is written via its equivalent in the form of maximum of the Lagrange function. Then minimax of this function is the object dual to the imitial one. For the problems of linear programming this leads to classical duality. To clarify this, we introduce the scheme

$$L: \max\{(c,x) \mid Ax \le b, \ x \ge 0\} \xrightarrow{\sim}$$

$$\xrightarrow{\sim} \max_{x \ge 0} \min_{u \ge 0}[(c,x) - (u, Ax - b)] = \max_{x \ge 0} \min_{u \ge 0}[(b,u) + (c - A^T u, x)],$$

$$\min_{u \ge 0} \max_{x \ge 0}[(b,u) + (c - A^T u, x)] \xrightarrow{\sim} L^*: \min\{(b,u) \mid A^T u \ge c, \ c \ge 0\};$$

here the symbol $\xrightarrow{\sim}$ denotes the equivalent transfer.

It occurs that, as in the classical sense, maximin of the disfunctive Lagrange function, i.e.

$$\max_{x \ge 0} \min_{u \ge 0} L_\cup(x, u)$$

is equivalent to the initial problem (which is (49.1)), it is denoted by the symbol P. The minimax of this function, i.e., $\min_{u \ge 0} \max_{x \ge 0} L_\cup(x, u)$ is equivalent to problem (49.6) which we shall show below. Problems (49.1) and (49.6) admit the equivalent rewriting

$$\max_{(j)} \max\{(c, x_j) \mid x_j \in M_j\}, \tag{49.8}$$

$$\max_{(j)} \min\{(c^j, u_j) \mid u_j \in M_j^*\}, \tag{49.9}$$

here

$$M_j = \{x_j \ge 0 \mid A_j x_j \le b^j\}, \qquad M_j^* = \{u_j \ge 0 \mid A_j^T x_j \ge c\}.$$

Before we establish the equivalence

$$\max_{x \ge 0} \min_{u \ge 0} L_\cup(x, u) \sim (49.1) \quad [\sim (49.8)], \tag{49.10}$$

$$\min_{u \ge 0} \max_{x \ge 0} L_\cup(x, u) \sim (49.6) \quad [\sim (49.9)] \tag{49.11}$$

we prove the auxiliary statement. Consider the problem

$$\min\left\{\max_{(j)}(b^j, u_j) \;\middle|\; u = [u_1, \ldots, u_m] \in M^* := \prod_{(j)} M_j^*\right\}. \qquad (49.12)$$

Lemma 49.1. *Problemd* (49.9) *and* (49.12) *are equivalent in the sense* opt (49.9) = opt (49.12) *and*

1) *if* $\bar{u}_{j_0} = \arg(49.9)$, *then* $\bar{u}_{j_0}$ *is the* j_0–*coordinate of a certain optimal vector* $\bar{u}$ *of problem* (49.12) *and conversely,*

2) *if* $\bar{u} = [\bar{u}_1, \ldots, \bar{u}_m] = \arg(49.12)$, $\max_{(j)}(b^j, \bar{u}_j) = (b^{j_0}, \bar{u}_{j_0})$, *then* $\bar{u}_{j_0} \arg(49.9)$.

Proof. We suppose that $\bar{t}_j := \operatorname{opt} L_j \; (= \operatorname{opt} L_j^*)$, $\bar{t} := \max_{(j)} \bar{t}_j = \bar{t}_{j_0}$. Evidently, $\bar{t} = \operatorname{opt}(49.9)$. Problem (49.12) can be rewritten in the equivalent form

$$\max\{\, t \mid (b^j, u_j) \le t, \; \forall j, \; u \in M^* \,\}. \qquad (49.13)$$

The value $\bar{t} \, (= \operatorname{opt}(49.9))$ is nondecreasing in (49.13) as it is nondecreasing in the problem

$$\min\{\, t \mid (b^{j_0}, u_{j_0}) \le t, \quad u_{j_0} \in M_{j_0}^* \,\}.$$

This gives us the equality opt (49.9) = opt (49.12). The properties 1) and 2) from the formulation of the lemma follows from the construction of the pair problems in question. □

Theorem 49.5. *Let problem* (49.1) *be solvable and completely regular, i.e.,* $M_j \ne \varnothing$, $\forall j$. *Then*

1. $$\max_{x \ge 0} \min_{u \ge 0} L_\cup(x, u) = \operatorname{opt}(49.1) \quad (\sim (49.8)), \qquad (49.14)$$

2. $$\min_{u \ge 0} \max_{x \ge 0} L_\cup(x, u) = \operatorname{opt}(49.6) \quad (\sim (49.12)). \qquad (49.15)$$

Proof. 1. We consider the internal operation in the left-hand side of (49.14):

$$\min_{u \geq 0} \left[(c, x) - \min_{(j)} (u_j, A_j(x) - b^j) \right] = \begin{cases} (c, x), & \text{if } \exists j' : A_{j'} x \leq b^{j'}, \\ -\infty, & \text{otherwise.} \end{cases}$$

This equality is clear. It yields

$$\max_{x \geq 0} \min_{u \geq 0} L_{\cup}(x, u) = \max \left\{ (c, x) \mid x \in \bigcup_{(j)} M_j \right\} = \text{opt} \ (49.1).$$

2. We write the internal operation in the left-hand side of (49.15) transforming the function $L_{\cup}(x, u)$:

$$\max_{x \geq 0} L_{\cup}(x, u) = \max_{x \geq 0} \left\{ \max_{(j)} \left[(b^j, u_j) + (c - A_j^T u_j, x) \right] \right\}.$$

Hence it follows

$$\max_{x \geq 0} L_{\cup}(x, u) = \begin{cases} \max_{(j)} (b^j, u_j), & \text{if } \forall j : A_j^T u_j \geq c, \\ +\infty, & \text{otherwise.} \end{cases}$$

This relation yields

$$\min_{u \geq 0} \max_{x \geq 0} L_{\cup}(x, u) = \min_{u \geq 0} \left\{ \max_{(j)} (b^j, u_j) \mid u = [u_1, \dots, u_m] \subset \prod_{(j)} M_j^* \right\}.$$

The problem standing in the right-hand side of thsi equality, by Lemma 49.1, is equivalent to problem (49.9), i.e., to problem (49.6) dual to (49.1). The theorem is proved. □

Remark 49.2. Indeed the equality from part 1 of Theorem 49.5 holds for general setting of problem of disjunctive programming (45.4) and its disjunctive Lagrange function (45.6). This can be varified analogously.

49.4. Symmetric setting of the problems (49.1) and (49.6) that are in the relation of duality

Problem (49.1) to which can be reduced each problem of piecewise linear programming falls into a finite number of parallel problems

$$L_j : \ \max\{(c, x) \mid A_j x \leq b^j, \ x \geq 0\}, \quad j = 1, \dots, m\}$$

whose optimal values form the optimal value of problem (49.1): opt (49.1) $=$ $\max_{(j)}$ opt L_j. Evident generalization of this setting is the problem

$$\max_{(j)} \max\{(c_j, x_j) \mid A_j x \le b^j, \ x \ge 0\},$$

which can be written in the equivalent form:

$$\max \left\{ f(x) := \max_{(j)}(c_j, x_j) \mid A_j x_j \le b^j, \ x_j \ge 0, \ j = 1, \ldots, m \right\}. \quad (49.16)$$

Following the scheme applied above, the dual problem will be as follows:

$$\min \left\{ f^*(u) := \max_{(j)}(b_j, u_j) \mid A_j^T u_j \ge c^j, \ u_j \ge 0, \ j = 1, \ldots, m \right\}. \quad (49.17)$$

These two problems (if the problems L_j, $\forall j$ are solvable) are connected by classical duality relations

1. $f(\bar{x}) \le f^*(\bar{u})$ for all admissible $\bar{x} = [\bar{x}_1, \ldots]$ and $\bar{u} = [\bar{u}_1, \ldots]$;

2. opt (49.16) $=$ opt (49.17).

The following pair problems which corresponds to that in (49.16) instead of $\max_{(j)}$ we take $\min_{(j)}$ also is mutually dual:

$$\max \left\{ \min_{(j)}(c_j, x_j) \mid A_j x_j \le b^j, \ x_j \ge 0, \ j = 1, \ldots, m \right\},$$

$$\min \left\{ \min_{(j)}(b_j, u_j) \mid A_j^T u_j \ge c^j, \ u_j \ge 0, \ j = 1, \ldots, m \right\}.$$

Note that if we consider the duality in piecewise linear programming then the following operation

$$\max \left\{ \min_{(j)} (\max) f_j(x_j) \ \middle| \ x = [x_1, \ldots] \in \prod_{(j)} M_j \right\},$$

where M_j is the space of the variable x_j, is the independent operation. This type of operation is used for the initial problem of σ–piecewise programming and for the dual problem.

Note also that the property $f(x) \le f^*(u)$ for admissible x and u allows us to reduce the solutions of problems (49.1) and (49.6) to solution of the system of piecewise linear inequalities

$$
\begin{cases}
(c, x) \geq \max_{(j)}(b^j, u_j), \\
\min_{(j)} |A_j x - b^j|_{\max} \leq 0, \\
A_j u_j \geq c, \\
x \geq 0, \quad u_j \geq 0, \quad j = 1, \ldots, m.
\end{cases}
\tag{49.18}
$$

Connection between problems (49.1), (49.6), and (49.18) is evident:

$$
\text{Arg}\,(49.18) = \text{Arg}\,(49.1) \times \text{Arg}\,(49.6),
$$

where the symbol Arg (49.18) denotes the set of solutions of system (49.18) which is the analog of the system $(c, x) \geq (b, u)$, $Ax \leq b$, $A^T u \geq c$, $x \geq 0$, $u \geq 0$ for the LP problem

$$
\max\{(c, x) \mid Ax \leq b, \ x \geq 0\}.
$$

50. DUALITY FOR IMPROPER PROBLEMS OF PIECE-WISE LINEAR PROGRAMMING

We shall write now the generalization of problems (49.1) and (49.6) in form (49.16) and (49.17) with the change of numerating indices. Namely, we set

$$
L_k : \ \max\{(c_k, x_k) \mid A_k x_k \leq b_k, \ x_k \geq 0\},
$$
$$
L_k^* : \ \min\{(b_k, u_k) \mid A_k^T u_k \geq c_k, \ u_k \geq 0\},
$$
$$
f(x) := \max_{(k)}(c_k, x_k), \qquad f^*(u) := \max_{(k)}(b_k, u_k),
$$
$$
M_k := \{\, x_k \mid A_k x_k \leq b_k, \ x_k \geq 0\},
$$
$$
M_k^* := \{\, u_k \mid A_k^T u_k \geq c_k, \ u_k \geq 0\},
$$
$$
k = 1, \ldots, \bar{k}.
$$

Problems (49.16) and (49.17) can be rewritten in symmetric form

$$
L : \ \max\left\{ f(x) \ \middle| \ x = [x_1, \ldots] \in \prod_{(k)} M_k \right\},
\tag{50.1}
$$

$$
L^\circledast : \ \min\left\{ f^*(u) \ \middle| \ u = [u_1, \ldots] \in \prod_{(k)} M_k^* \right\}.
\tag{50.2}
$$

We consider the case of improperness (unsolvability) of problems L and $L^{\circledast}$ with the duality position. As these problems formally fall into the collection of problems $\{L_k\}$ and $\{L_k^*\}$, then we can apply the duality scheme for improper LP problems to the pairs of mutually dual problems $\{L_k, L_k^*\}$ independently of their solvability (see Section 25):

$$
\begin{array}{ccc}
L_k & \xrightarrow{\ \pi\ } & C_k \\[4pt]
\Updownarrow & & \Updownarrow \\[4pt]
L_k^* & \xrightarrow{\ \pi\ } & C_k^{\#}
\end{array}
$$

Following this scheme, the problems C_k and $C_k^{\#}$ are solvable and $\operatorname{opt} C_k = \operatorname{opt} C_k^{\#}$. The problems C_k and $C_k^{\#}$ (following Eremin *et al.*, 1983; Section 6) have the form

$$
\max\left\{ (c_k, x_k) - \sum_{j=1}^{m_k} R_j^k \|(A_j^k x_k - b_k^j)^+\|_{p_{jk}} \ \middle| \right.
$$
$$
\left. \middle| \ A_0^k x_k \le b_k^0, \quad \|x_k^i\|_{q_{ik}} \le r_i^k, \quad i = 1, \ldots, n_k, \quad x_k \ge 0 \right\}, \qquad (50.3)
$$

$$
\min\left\{ (b_k, u_k) - \sum_{i=1}^{n_k} r_i^k \|(c_k^i - {B_i^k}^T u_k)^+\|_{q_{ik}}^* \ \middle| \right.
$$
$$
\left. \middle| \ {B_0^k}^T u_k \ge c_k^0, \quad \|u_k^j\|_{p_{jk}}^* \le R_j^k, \quad j = 1, \ldots, m_k, \quad u_k \ge 0 \right\}, \qquad (50.4)
$$

Here $\{A_j^k\}_{j=0}^{m_k}$ and $\{B_i^k\}_{i=0}^{n_k}$ are arbitrary cuts of the matrix A_k onto horizontal and vertical submatrices; $\{c_k^i, x_k^i\}$ and $\{b_k^j, u_k^j\}$ are the correspondent cuts of the vectors $\{c_k, x_k\}$ and $\{b_k, u_k\}$; $\{\|\cdot\|_{p_{jk}}$ and $\{\|\cdot\|_{q_{ik}}$ are the collections of norms in correpondent spaces; $\{R_j^k\}$ and $\{r_i^k\}$ are non-negative parameters. Subsystems $A_0^k x_k \le b_k^0$, $x_k \ge 0$; and ${B_0^k}^T u_k \ge c_k^0$, $u_k \ge 0$ in (50.3) and (50.4) are assumed to be compatible. For certain r_i^k and R_j^k the systems of restrictions in the problems C_k and $C_k^{\#}$ will be compatible also. The sense of the duality theorem for C_k and $C_k^{\#}$ is in the equality $\operatorname{opt} C_k = \operatorname{opt} C_k^{\#}$ (Theorem 25.2).

In the basis of duality for L and $L^{\circledast}$ we may set the scheme

$$L \longrightarrow \{L_k\} \xrightarrow{\pi} \{C_k\} \longrightarrow C$$

$$\Updownarrow \qquad \Updownarrow (*) \qquad \Updownarrow (\#) \qquad \Updownarrow (\#)$$

$$L^{\circledast} \longrightarrow \{L_k^*\} \xrightarrow{\pi} \{C_k^{\#}\} \longrightarrow C^{\#}$$

where formation of the problems C and $C^{\#}$ in this scheme is similar to those of formation (50.1) and (50.2) for the problems $\{L_k\}$ and $\{L_k^*\}$.

We suppose that $g_k(x_k)$ and $g_k^{\#}(u_k)$ are the goal functions in problems (50.3) and (50.4); N_k and $N_k^{\#}$ are admissible sets for these problems;

$$g(x) := \max_{(k)} g_k(x_k), \qquad g^{\#}(x) := \max_{(k)} g_k^{\#}(u_k).$$

We form the problems

$$C : \sup \{g(x) \mid x = [x_1, \ldots] \in N := \textstyle\prod N_k\},$$
$$C^{\#} : \inf \{g^{\#}(x) \mid u = [u_1, \ldots] \in N^{\#} := \textstyle\prod N_k^{\#}\}.$$

Theorem 50.1. *Let the conditions hold:*

1. *The parameters r_i^k and R_j^k are chosen so that the systems of restrictions in (50.3) and (50.4) are compatible for rigorous inequalities $\|x_k^i\|_{q_{ik}} < r_i^k$ $(i = 1, \ldots, n_k)$ $\|u_k^j\|_{p_{jk}}^* < R_j^k$ $(j = 1, \ldots, m_k)$.*

2. *All the norms used for the problems $\{C_k\}$ and $\{C_k^{\#}\}$ are monotone.*

3. *The operation sup in C is accessible. (its finiteness follows from the condition 1 due to the property $g(x) \le g^{\#}(u)$, $\forall [x, u] \in N \times N^{\#}$).*

Then the problem $C^{\#}$ is solvable and opt $C = $ opt $C^{\#}$.

This theorem follows from the basis duality theorem for $IPLP$ which we had mentioned and the considerations similar to the proof of Theorem 49.4.

Remark 50.1. Theorem 50.1 is rather general, it contains a lot of parameters with the wide range of their change (dividing of restriction system onto subsystems, the choice of norms, the choice of the parameters R_j^k and r_i^k). Therefore, there exists a lot of particular formulations which have their own interest. We shall not consider them here since we shall go off the course of our chapter.

51. THE METHOD OF EXACT PENALTY FUNCTIONS FOR THE PROBLEM OF PIECEWISE LINEAR PROGRAMMING

We consider the general problem of piecewise linear programming in canonical setting (49.1), i.e.,

$$P_\cup : \ \max\big\{(c,x) \ \big| \ \min_{j=1,\dots,m} |A_j x - b^j|_{\max} \leq 0, \ x \geq 0\big\} \qquad (51.1)$$

and reduce this problem to equivalent problem of similar type but without the base restriction in (51.1). We consider parallel with the problem $P_\cup$ the k-problem

$$\sup_{x \geq 0} \big[(c,x) - \min_{(j)}(R_j, (A_j x - b^j)^+)\big], \qquad (51.2)$$

where R_j is a non-negative parameter vector of the dimension m_j; i.e., m_j is the number of inequalities in the system $A_j x - b^j \leq 0$.

As the previous paragraph, we shall use the following notations:

$$L_j : \ \max\{(c,x) \mid A_j x \leq b^j, \ x \geq 0\},$$

L_j^* is dual to L_j, $\bar{u}_j = \arg L_j^*$, $\bar{u} = [\bar{u}_1, \dots, \bar{u}_m]$, $M_j = \{x \geq 0 \mid A_j x \leq b^j\}$.

Theorem 51.1. *We suppose that* (51.1) *is solvable,* $M_j \neq \varnothing$, $\forall j$; $\bar{u}_j \in \text{Arg}\, L_j^*$. *If* $R_j \geq R_0 \bar{u}_j$, $R_0 > 1$, *then the optimal values and the optimal sets of problems* (51.1) *and* (51.2) *coincide, i.e.,*

$$\text{opt}\,(51.1) = \text{opt}\,(51.2), \qquad (51.3)$$
$$\text{Arg}\,(51.1) = \text{Arg}\,(51.2). \qquad (51.4)$$

Proof. We denote the goal function in (51.2) by $\Phi_R(x)$; and the term subtracted from (c,x) – by $\Phi_0(x)$. First, we shall prove equality (51.3). For $\bar{x} \in \text{Arg}\,(51.1)$ we obtain

$$\Phi_R(\bar{x}) = (c, \bar{x}) = \text{opt}\,(51.1);$$

therefore,

$$\text{opt}\,(51.2) = \sup_{x \geq 0} \Phi_R(x) \geq \text{opt}\,(51.1).$$

We establish now the inverse inequality. By Theorem 47.2, we have

$$(c, x) - \min_{(j)}(\bar{u}_j, A_j x - b^j) \leq (c, \bar{x}), \qquad \forall x \geq 0.$$

Taking into account this inequality, we obtain the estimate for $\Phi_R(x)$ for $x \geq 0$:

$$\Phi_R(x) \leq (c, \bar{x}) + \min_{(j)}(\bar{u}_j, A_j x - b^j) - \Phi_0(x)$$

$$\leq \operatorname{opt}(51.1) + \min_{(j)}(\bar{u}_j, (A_j x - b^j)^+) - \Phi_0(x)$$

$$\leq \operatorname{opt}(51.1) + \frac{1}{R_0} \min_{(j)}(R_j, (A_j x - b^j)^+) - \Phi_0(x)$$

$$= \operatorname{opt}(51.1) - \frac{R_0 - 1}{R_0} \min_{(j)}(R_j, (A_j x - b^j)^+)$$

$$\leq \operatorname{opt}(51.1). \tag{51.5}$$

Hence it follows that $\sup_{x \geq 0} \Phi_R(x) \leq \operatorname{opt}(51.1)$. Thereby, equality (51.3) is proved. From this inequality, in particular, the inclusion $\operatorname{Arg}(51.1) \subset \operatorname{Arg}(51.2)$ follows. This allows us to write max in problem (51.2) instead of sup. We shall show now the inverse inclusion. Suppose $\bar{x} \in \operatorname{Arg}(51.2)$. Following (51.5), we have

$$\operatorname{opt}(51.1) = \Phi_R(\bar{x}) \leq \operatorname{opt}(51.1) - \frac{R_0 - 1}{R_0} \min_{(j)}(R_j, (A_j \bar{x} - b^j)^+).$$

Hence it follows

$$\min_{(j)}(R_j, (A_j \bar{x} - b^j)^+) = 0;$$

therefore, $\exists j_0 : (R_{j_0}, (A_{j_0} \bar{x} - b^{j_0})^+) = 0$. Taking into account $R_{j_0} > 0$ this yields $A_{j_0} \bar{x} \leq b^{j_0}$. Therefore, $\bar{x} \in M_{j_0} \subset M = \bigcup_{j=1}^{m} M_j$. If $\bar{x}$ is admissible for problem (51.1) and $(c, \bar{x}) = \operatorname{opt}(51.1)$ then $\bar{x} \in \operatorname{Arg}(51.1)$; therefore, $\operatorname{Arg}(51.2) \subset \operatorname{Arg}(51.1)$. So, (51.4) is established also. $\square$

This proof can be realized and for more general situation; namely, for problem (47.1) and for its equivalent reduction to the problem

$$\sup_{x \geq 0} \left[f(x) - \min_{(j)}(R_j, F_j^+(x)) \right]. \tag{51.6}$$

Theorem 51.2. *Let each problem* $\max\{f(x) \mid F_j(x) \le 0, \ x \ge 0\}$ *have the saddle point* $[\bar{x}_j, \bar{u}_j]$. *Then, for* $R_j > R_0 \bar{u}_j$, $R_0 > 1$ *problems* (47.1) *and* (51.6) *are equivalent in the sense of coincidence of their optimal values and optimal sets.*

Really, following Theorem 47.2, we may consider the inequality

$$f(x) \le f(\bar{x}) + \min_{(j)}(\bar{u}_j, F_j(x)), \qquad \forall x \ge 0$$

and then repeat all following calculations (51.5).

52. QUESTIONS OF POLYHEDRAL SEPARABILITY

The problem of separation of the two nonintersecting sets M and N from a certain space $\mathbb{X}$ by means of the function $f(x)$ from a certain class $\mathcal{F}_0$ is called *the problem of discrimination of sets*. The function $f(x)$ in this case is called *the discriminant function*. Separation of sets in this case looks as follows:

$$f(x) > 0, \quad \forall x \in M; \qquad f(y) < 0, \quad \forall y \in N. \tag{52.1}$$

We can consider not rigorous separation, when the inequalities in (52.1) are not strict.

If M and N are convex polyhedrons and $\mathcal{F}_0$ is the class of affine functions, then we say about *linear discrimination*. The problem of linear discrimination is one of most important problems in pattern recognition (PR). One of the formal basis models of PR is the system of strict linear homogeneous inequalities. We shall show this.

Suppose $\mathcal{G}$ and $\mathcal{L}$ are certain patterns, $A = \{a_j\} \subset \mathbb{R}^n$ and $B = \{b_i\} \subset \mathbb{R}^n$ are formalized finite samples from these patterns. If $\{f_s(x)\}_1^k$ is a certain basis collection of functions (generally speaking, arbitrary) then the selected function can be searched in the form $f(x) = \sum_{s=1}^{k} z_s f_s(x)$, where $\{z_s\}$ are numerical coefficients. The separation property is as follows:

$$\sum_{s=1}^{k} z_s f_s(a_j) > 0, \quad \forall j; \qquad \sum_{s=1}^{k} z_s f_s(b_i) < 0, \quad \forall i,$$

or, in the matrix form,

$$\bar{A}z > 0, \qquad \bar{B}z < 0, \tag{52.2}$$

where $\bar{A} := [a_{js}], \ a_{js} := f_j(a_s); \ \bar{B} := [b_{is}], \ b_{is} := f_s(b_i)$.

If $\bar{z} = [\bar{z}_1, \ldots, \bar{z}_k]$ is a certain solution of system (52.2), then the function $f(x) = \sum_{s=1}^{k} \bar{z}_s f_s(x)$ will be strictly separating for sets A and B. The discriminant function $f(x)$ realized the rule of relation (by the property of membership to one of the patterns $\mathcal{G}$ and $\mathcal{L}$) for the arbitrary vector y following the law

$$y \in \mathcal{G}, \quad \text{if} \ f(y) > 0;$$
$$y \in \mathcal{L}, \quad \text{if} \ f(y) < 0.$$

The function $f(x)$ realized *the resolving rule*. System (52.2) may be both compatible and incompatible. If the system is incompatible, we have the generalization of notion of solution, as a finite collection of vectors $\{c_l\} \subset \mathbb{R}^n$ (which is called *committe solution*) such that each of inequalities of system (52.2) is satisfied more than half of vectors of this collection. Committee technology and its use in pattern recognition is the important and well-developed direction in pattern recognition (Mazurov, 1990).

If M and N are polyhedrons, and $M \cap N = \varnothing$, then these sets are strictly separated by the affine function. Further, we shall deal with separation of arbitrary noninteresting polyhedrons sets by piecewise linear function (k-function).

So, we suppose that $M = \bigcup_{(j)} M_j$, $N = \bigcup_{(i)} N_i$, $\{M_j\}$ and $\{N_i\}$ – are finite collections of convex polyhedrons, where $M \cap N = \varnothing$.

Theorem 52.1. *Polyhedral sets M and N with empty intersection are strictly separated by the k-function, i.e. by the function of form (48.2):*

$$f(x) = \min_{j \in \overline{1,m}} |A_j x - b^j|_{\max}. \tag{52.3}$$

Proof. The polyhedral set can be setted by one k-inequality: if $M_j = \{x \mid A_j x \le b^j\}$ and $N_i = \{x \mid B_i x \ge d^i\}$, then

$$M = \{x \mid f(x) \le 0\},$$

where $f(x)$ is a function of form (52.3);

$$N = \{x \mid g(x) \ge 0\},$$

where $g(x) = \max_{(j)} |B_i x - d^i|_{\min}$. Taking into account the relations $(\mathbb{X} \setminus M) \supset N$, $(\mathbb{X} \setminus N) \supset M$ for each $\alpha > 0$ and $\beta > 0$ we shall have

$$\alpha f(x) + \beta g(x) < 0, \quad \forall x \in M;$$

$$\alpha f(y) + \beta g(y) > 0, \quad \forall y \in N.$$

Thereby, we have constructed the function $f_{\alpha,\beta}(x) := \alpha f(x) + \beta g(x)$ dependent on the numerical parameters $\alpha > 0$ and $\beta > 0$ which strictly separates the polyhedral sets M and N. $\quad\square$

Corollary 52.1. *Let $\mathbb{R}^n \supset \{a_j\}_{j\in J}$ and $\mathbb{R}^n \supset \{b_i\}_{i\in I}$ be finite collections of points from $\mathbb{R}^n$. In this case, $\{a_j\} \cap \{b_i\} = \varnothing$. Then there exists a k-function $f(x)$ strictly separating these sets:*

$$f(a_j) < 0, \quad \forall j \in J; \qquad f(b_i) > 0, \quad \forall i \in I.$$

Really as the point from $\mathbb{R}^n$ can be setted by a finite system of linear inequalities (if $\bar{x} = [\bar{x}_1, \ldots, \bar{x}_n]$, then $\bar{x} = \{x \mid \bar{x} \leq x \leq \bar{x}\}$), then this corollary is a particular case of Theorem 52.1.

Indeed, the space where the point are taken from may be arbitrary. Reduction to a finite dimensional arithmetic space is trivial: it suffices to take a subspace $\mathbb{E}_k$ of the initial space stretched on the collection $\{a_j\}_{j\in J} \cup \{b_i\}_{i\in I}$ and choose a certain basis in it. In this basis the vectors a_j and b_i will be represented by the points of a finite dimensional arithmetic space. If we take the direct sum $\mathbb{X} = \mathbb{E}_k + \mathbb{H}$, then the k–function $f(x)$ separated the mentioned collections of points in $\mathbb{E}_k$ can be extended to the k–function $\bar{f}(x)$ defined in the whole $\mathbb{X}$ by the rule:

if $x \in \mathbb{X}$ and $x = \bar{x} + h$, $\bar{x} \in \mathbb{E}_k$, $h \in \mathbb{H}$, then $\bar{f}(x) = f(\bar{x})$.

Appendix.

Elements of convex analysis and convex programming

The information given in Appendix is beyond the basic contents of the book; however, some problems of linear programming included in the monograph, for example, the problems on marginal values, stability, require for their consideration the elements of nonlinear (convex) analysis. Therefore, we give here some definitions and properties which we have used in the basic text.

The statements cited below (Theorems 1–14) can be found in the books (Eremin and Astaf'ev, 1976) and (Rockafellar, 1970)

A1. CONVEX SETS AND CONVEX FUNCTIONS

A set $M \subset \mathbb{R}^n$ is called *convex* if

$$\forall x, y \in M \implies \alpha x + (1 - \alpha)y \in M, \qquad \forall \alpha \in [0, 1].$$

A convex set M is called *solid* if $\exists p \in M, \exists \varepsilon > 0 : \{x \mid \|x - p\| \leq \varepsilon\} \subset M$.

The point p in this case is called *interior point*. The set of all interior points of the set M is called *interior*. The interior of the set M we shall denote as M^0.

A function $f(x)$ setted in a convex set M is called *convex* if for each $x, y \in M$ and $\alpha \in [0, 1]$ the inequality

$$f(\alpha x + (1 - \alpha)y) \leq \alpha f(x) + (1 - \alpha)f(y) \qquad (1)$$

holds. If for $x \neq y$ the inequality (1) is rigorous, then $f(x)$ is called *strictly convex*.

A function $f(x)$ is called *concave* (*strictly concave*) if $-f(x)$ is convex (strictly convex).

Convex functions have the following evident properties

1) a function $f(x)$ is convex $\Longrightarrow$ the set $\{x \in M : f(x) \leq \alpha\}$ is convex (or empty) for each $\alpha \in \mathbb{R}$;

2) functions $\{f_j(x)\}_1^m$ are convex $\Longrightarrow$ the set $\sum_{j=1}^m \alpha_j f_j(x)$ is convex for each $\alpha_j \geq 0$, $j = 1, \ldots, m$;

3) if $f(x)$ is convex, then

$$f\left(\sum_{j=1}^m \alpha_j x_j\right) \leq \sum_{j=1}^m \alpha_j f(x_j)$$

for each finite collections $\{x_j\}_1^m \subset M$; here $\alpha_j \geq 0$, $\sum \sum_{j=1}^m \alpha_j = 1$.

Theorem 1. *A convex function $f(x)$ gives in a convex set $M \subset \mathbb{R}^n$ is continuous in each point $x \in M^0$.*

Theorem 2. *If a convex function $f(x)$ is differentiable in a certain extension of a convex set M (for example, in $M^\varepsilon := \{x \mid |x - M| \leq \varepsilon\}$, $\varepsilon > 0$) then it is continuously differentiable in M (the gradient $\nabla f(x) = \left[\frac{\partial f(x)}{\partial x_1}, \ldots, \frac{\partial f(x)}{\partial x_n}\right]$ is continuous).*

Theorem 3. *Let a function $f(x)$ be differentiable in a certain extension of a convex set M; then it is convex in M if and only if*

$$(\nabla f(x), y - x) \leq f(y) - f(x), \qquad \forall x, y \in M. \qquad (2)$$

A2. SUBDIFFERENTIABILITY OF CONVEX FUNCTIONS

Relation (2) can be put in the basis of definition of generalized gradient (subgradient).

A vector $h \in \mathbb{R}^n$ is called *subgradient* of a function $f(x)$ in a point $x \in M$ if the inequality

$$(h, y - x) \leq f(y) - f(x) \tag{3}$$

holds for all $y \in M$.

The set of all cubgradients in the point x is denoted by $\partial f(x)$ and is called *subdifferential*.

Theorem 4. *Let $f(x)$ be a convex function given in a convex set M with nonempty interior M^0. For $x \in M^0$ the following properties hold.*

1) $\partial f(x) \neq \varnothing$;

2) $\partial f(x)$ is convex and bounded;

3) $\partial f(x)$ is closed.

Some rules of subdifferentiation are given in the next theorem.

Theorem 5. *The following properties hold:*

1) $\partial(\beta f(x)) = \beta \partial f(x)$ for each $\beta \geq 0$ and $x \in M^0$;

2) $\partial(f(x) + \beta) = \partial f(x)$ for each $\beta \in \mathbb{R}$ and $x \in M^0$;

3) if $c \in \mathbb{R}^n$ then $\partial(f(x) + (c, x)) = \partial f(x) + c$ for $x \in M^0$;

4) if $f(x)$ and $g(x)$ are convex functions given in M, then

$$\partial(f(x) + g(x)) = \partial f(x) + \partial g(x) \quad \text{for} \quad x \in M^0.$$

The following remark is useful.

The inclusion $0 \in \partial f(p)$ is the necessary and sufficient condition that p is a point of global minimum of the function $f(x)$.

The derivative in the direction l of a function $f(x)$ in a point x_0 is the value

$$\frac{\partial f(x)}{\partial l}\bigg|_{x=x_0} := \lim_{t \to +0} \frac{f(x_0 + tl) - f(x_0)}{t}$$

if the limit exists. If $f(x)$ is differentiable in the point x_0, then $\frac{\partial f(x)}{\partial l} = (\nabla f(x_0), l)$.

Theorem 6. *Let $f(x)$ be a convex function setted in a convex set M and $M^0 \neq \varnothing$. For $x_0 \in M^0$ the following formula holds*

$$\frac{\partial f(x)}{\partial l}\bigg|_{x=x_0} = \max_{h \in \partial f(x_0)} (h, l).$$

A3.   THE PROBLEM OF CONVEX PROGRAMMING

The problem of convex programming (CP) is the following problem:

$$\max \{ f(x) \mid f_j(x) \leq 0, \ j = 1, \ldots, m, \ x \in M_0 \}, \tag{4}$$

where $-f(x), f_1(x), \ldots, f_m(x)$ are convex functions; M_0 is a convex set. The admissible set of this problem $M = \{ x \in M_0 \mid f_j(x) \leq 0, \ j = 1, \ldots, m \}$ is convex (or empty).

Theorem 7. *The notions of local and global optimum for problem* (4) *coincide.*

We introduce for problem (4) the condition of *R-regularity*:

$$\exists p \in M^0 : \ f_j(p) < 0, \quad j = 1, \ldots, m.$$

The condition of regularity for CP problem in bibliography is met in various forms. The essence of these conditions is in providing the Kun-Takker theorem (here, Theorem 8). Weakening R-regularity condition may be done as follows:

If all the restrictions of the problem are written in the form of inequality system (among them the set M_0 in (4)), then there exists an admissible vector p such that nonlinear restrictions for this p hold strictly.

In this case we shall say about R_0–*regularity*. In particular, for linear problems of optimization, the R_0–regularity condition will hold automatically.

Suppose now that $F(x, u) = f(x) - \sum_{j=1}^{m} u_j f_j(x)$ is the Lagrange function correspondent to problem (4).

Theorem 8 [Kuhn−Tucker]. *If CP problem (4) is solvable, suppose in the point $\bar{x}$ and is R_0-regular, then $\exists \bar{u} \geq 0$, $\bar{u} \in \mathbb{R}^m$ such that*

$$F(x, \bar{u}) \underset{\forall x \in M_0}{\leq} F(\bar{x}, \bar{u}) \underset{\forall u \geq 0}{\leq} F(\bar{x}, u), \tag{5}$$

where $\bar{u}_j f_j(\bar{x}) = 0$, $j = 1, \ldots, m$.

The vector $[\bar{x}, \bar{u}]$ satisfying (5) is called the saddle point for the Lagrange function $F(x, u)$.

Note that condition (5) is equivalent to the condition

$$\max_{x \in M_0} \min_{u \geq 0} F(x, u) = \min_{u \geq 0} \max_{x \in M_0} F(x, u). \tag{6}$$

Theorem 9. *If $[\bar{x}, \bar{u}]$ is the saddle point for the function $F(x, u)$ correspondent to problem (4), then $\bar{x}$ is the optimal solution for (4) without any assumptions relatively the functions entering this problem and the set M_0. In this case, $\bar{u}_j f_j(\bar{x}) = 0$, $j = 1, \ldots, m$.*

We define the problem

$$\min_{u \geq 0} \max_{x \in M_0} F(x, u) \tag{7}$$

as dual to (4). Then the Kun-Takker theorem can be reformulated in the form of direct duality theorem.

Theorem 10. *If for initial problem (4) the conditions of Theorem 8 hold, then dual problem (7) is solvable. In this case optimal values of direct and dual problems coincide and $\bar{u}_j f_j(\bar{x}) = 0$, $j = 1, \ldots, m$.*

We write problem (4) for $M_0 = \mathbb{R}^n$:

$$\max\{f(x) \mid f_j(x) \leq 0, \; j = 1, \ldots, m\}. \tag{8}$$

Theorem 11. *Let* (8) *be a CP problem and* $\bar{x} \in M = \{ x \mid f_j(x) \leq 0,$ $j = 1, \ldots, m \}$. *The condition*

$$h = \sum_{j \in J(\bar{x})} u_j h_j, \tag{9}$$

where $h \in \partial f(\bar{x})$, $h_j \in \partial f_j(\bar{x})$, $u_j \geq 0$, $j \in J(\bar{x}) := \{ j \mid f_j(\bar{x}) = 0 \}$ *is sufficient for* $\bar{x} \in \text{Arg}\,(8)$. *If for* (8) *the* R_0*-regularity condition holds, then this condition is also necessary.*

We suppose now that problem (8) is smooth, i.e. the functions $\{ f(x)$, $f_j(x)$, $j = 1, \ldots, m \}$ are differentiable. Let $F(x, u) = f(x) - \sum_{j=1}^{m} u_j f_j(x)$ be its Lagrange function and

$$\min\{ f(x, u) \mid \nabla F(x, u) = 0, \ u \geq 0 \} \tag{8*}$$

be the problem dual by Vulf (Eremin and Astaf'ev, 1976).

Theorem 12 [direct duality theorem]. *Suppose CP problem* (8) *is solvable and* R_0*-regular. If* $\bar{x} \in \text{Arg}\,(8)$, *then there exists such a vector* $\bar{u} \in 0$ *that* $[\bar{x}, \bar{u}] \in \text{Arg}\,(8^*)$. *In this case,* $\sum_{j=1}^{m} \bar{u}_j f_j(\bar{x}) = 0$ *and* $\text{opt}\,(8) = \text{opt}\,(8^*)$.

Theorem 13 [converse duality theorem]. *Suppose problem* (8*) *is solvable. If* $[\bar{x}, \bar{u}] \in \text{Arg}\,(8^*)$ *and* $F(x, \bar{u})$ *is strictly concave in a certain neighborhood of the point* $\bar{x}$, *then CP problem* (8) *is solvable and* $\bar{x} \in \text{Arg}\,(8)$.

A4. THE THEOREM ON MARGINAL VALUES

We shall write now the parameter problem of convex programming

$$\max \{ f(x, y) \mid f_j(x, y) \leq 0, \ j = 1, \ldots, m \}, \tag{10}$$

here y is a parameter which can be considered as information component of problem (8). If it is the *LP* problem: $\max\{ (c, x) \mid Ax \leq b \}$, then we may set $y = [A, b, c]$. For a fixed y, (10) is a *CP* problem. Indeed, in theorem formulations considered behaviour of the function $\tilde{f}(y) := \text{opt}\,(10)$ in a neighborhood V of a fixed value y_0 of the parameter y, it suffices to suggest in problem (10) the convexity of the function only for $y \in V$. We set

$$F(x, u, y) = f(x, y) - \sum_{j=1}^{m} u_j f_j(x, y), \tag{11}$$

where $M(y)$ is the admissible set of problem (10); $\widetilde{M}(y) := \arg(10)$, $\widetilde{M}^*(y) := \{\bar{u} \geq 0 \mid \exists \bar{x} \in \widetilde{M}(y) : [\bar{x}, \bar{u}] \text{ is a saddle point of the Lagrange function (11)}\}$.

Theorem 14. *Suppose for problem* (10) *the following conditions hold:*

1. (10) *is a CP problem for each* $y \in V$;

2. *for* $(10)|_{y=y_0}$ *the R–regularity condition holds, i.e.,* $\exists p : f_j(p, y_0) < 0$, $j = 1, \ldots, m$;

3. $\widetilde{M}(y_0) = \{x_0\}$, $\widetilde{M}^*(y_0) = \{u_0\}$;

4. *the functions* $\{-f(x, y), f_1(x, y), \ldots, f_m(x, y)\}$ *are differentiable with respect to* y *in the point* y_0 *for each* x;

5. *for each* $\delta > 0$ *there exists* ε $(0 < \varepsilon \leq \delta)$ *such that* $\forall y : \|y - y_0\| \leq \varepsilon$, $\forall x \in \bigcup_{y \in V} M(y) \implies |f(x, y) - f(x, y_0)| \leq \delta$, $|f_j(x, y) - f_j(x, y_0)| \leq \delta$, $j = 1, \ldots, m$.

Then

$$\frac{\partial \tilde{f}(y_0)}{\partial l} = (F_y'(x_0, u_0, y)|_{y=y_0}, l). \tag{12}$$

Remark. If in (10) all the functions are linear, and y is the informative component of the problem, i.e., (10) is

$$\max\{(c, x) \mid Ax \leq b\}, \tag{13}$$

$y_0 = [A, b, c]$, then in Theorem 14 from all the conditions only the two remain: the regularity condition, i.e. $\exists p : Ax < b$ and the condition of uniqueness of solutions of problems (13) and (13*):

$$\min\{(b, u) \mid A^T u = c, \ u \geq 0\}. \tag{13*}$$

Formula (12), as it may be shown, attains the form

$$\frac{\partial \tilde{f}(A, b, c)}{\partial l} = (\Delta c, x_0) - (\Delta A \, x_0 - \Delta b, \, u_0), \tag{14}$$

where $l = [\Delta A, \Delta b, \Delta c]$.

A5. THE PENALTY FUNCTION METHOD FOR PROBLEMS OF NONLINEAR PROGRAMMING

We consider the problem of mathematical programming in the form

$$\max \left\{ f(x) \mid f_j(x) \leq 0, \ j = 1, \ldots, m, \ x \in M \subset \mathbb{X} \right\}, \tag{15}$$

where $\mathbb{X}$ is an arbitrary real linear space. The following reduced problem (see (Eremin, 1996)) will correspond to problem (15):

$$\sup_{x \in M} \left\{ f(x) - P\left(r_1 \| F_1^+(x) \|_1^{\alpha_1}, \ldots, r_k \| F_k^+(x) \|_k^{\alpha_k} \right) \right\}. \tag{16}$$

Here $F(x) = [F_1(x), \ldots, F_k(x)] = [f_1(x), \ldots, f_m(x)]$, $F_i(x) \in \mathbb{E}_{n_i}$, $\sum_{i=1}^k n_i = m$, $\| \cdot \|_i$ is an arbitrary norm of the space $\mathbb{E}_{n_i}$, $F_i^+(x)$ is the positive cut of the vector $F_i(x)$; $\{r_i\}_1^k$ and $\{\alpha_i\}_1^k$ are positive parameters; $P(z)$ is a function convex in $\mathbb{E}_k^+$ such that $P(0) = 0$, $P(z) > 0$ for $z \geq 0$, $z \neq 0$.

We define the saddle point $[\bar{x}, \bar{u}] \in M \times \mathbb{E}_k^+$ as follows:

$$F(x, \bar{u}) \underset{\forall x \in M}{\leq} F(\bar{x}, \bar{u}) \underset{\forall u \geq 0}{\leq} F(\bar{x}, u), \tag{17}$$

where $F(x, u) = f(x) - \sum_{j=1}^m u_j f_j(x)$ is the Lagrange function. Following the partition $[F_1(x), \ldots, F_k(x)]$ the vector $\bar{u}$ falls into the components $\bar{u}_i : \bar{u} = [\bar{u}_1, \ldots, \bar{u}_k]$. In case (17) we evidently have $(\bar{u}, F(\bar{x})) = 0$ and $\bar{x} \in \mathrm{Arg}\,(15)$.

Theorem 15. *Suppose the conditions hold*

1) for $F(x, u)$ there exists a saddle point $[\bar{x}, \bar{u}]$;

2) $\exists h = [h_1, \ldots, h_k] > 0 : h \in \partial P(0)$ is the subdifferential of the convex function $P(z)$ in the point $z = 0$;

3) $\alpha_i = 1$, $i = 1, \ldots, k$;

4) $r_i \geq \dfrac{\|\bar{u}_i\|_i^*}{h_i}$, $i = 1, \ldots, k$.

Then opt $(15) =$ opt (16). *If condition 4) holds with rigorous inequalities, then* Arg $(15) =$ Arg (16).

Theorem 16. *Let the condition 1) and 2) from Theorem 15 hold and 3) $\alpha_i > 1$, $i = 1, \ldots, k$. Then the estimate*

$$0 \leq \text{opt}\,(16) - \text{opt}\,(15) \leq \Delta(r)$$

holds, where

$$\Delta(r) = \sum_{i=1}^{k} \frac{\alpha_i - 1}{\alpha_i} (\|\bar{u}_i\|_i^*)^{\frac{\alpha_i}{\alpha_i - 1}} (\alpha_i h_i r_i)^{-\frac{1}{\alpha_i - 1}}.$$

Corollary. opt $(16) \rightarrow$ opt (15) *for $r_0 := \min_{(i)} r_i \rightarrow +\infty$.*

Remark. Examples of the functions satisfying condition 2) are the functions

$$P_1(z) = \sum_{i=1}^{k} |z_i|, \qquad P_2(z) = \max_{(i)} |z_i|.$$

A6. AN ESTIMATE OF DEVIATION WITH RESPECT TO THE ARGUMENT IN THE ASYMPTOTIC PENALTY METHOD

We write the initial problem

$$\max \left\{ f(x) - \alpha \|x - p\|^2 \mid f_j(x) \leq 0, \quad j = 1, \ldots, m, \quad x \in M \subset \mathbb{R}^n \right\} \quad (18)$$

and the problem with penalty

$$\max\left\{f(x) - \alpha\,\|x - p\|^2 - \sum_{j=1}^{m} R_j f_j^{+2}(x) \;\Big|\; x \in M\right\}, \qquad \alpha > 0. \qquad (19)$$

Suppose that $\bar{x}_\alpha \in \text{Arg}\,(18)$, $\bar{x}_R \in \text{Arg}\,(19)$.

Theorem 17. *Let problem* (18) *be solvable and satisfy the conditions that provide the saddle point* $[\bar{x}, \bar{u}]$ *for its Lagrange function. Then the following estimate holds*

$$\|\bar{x}_R - \bar{x}_\alpha\| \le \sqrt{\frac{1}{8\alpha} \sum_{j=1}^{m} \frac{\bar{u}_j^2}{R_j}}.$$

In particular, if we set $R_j = \alpha^{-3}$, $\forall j$, then the estimate attains the form

$$\|\bar{x}_R x_R - \bar{x}_\alpha\| \le \frac{\alpha\,\|\bar{u}\|}{2\sqrt{2}}.$$

NOTATIONS AND ABBREVIATIONS

$:=$ $(=:)$ —the left-hand (right-hand) side of the equality is the notation for the right-hand (left-hand) side;

$\mathbb{E}_n$ —the linear space of the vectors $x := [x_1, \ldots, x_n]^T$, where T denotes the transpose of a matrix or, in particular, of a vector;

$\mathbb{R}^n$ —is the space $\mathbb{E}_n$ alloted by the norm

$$\|x\| := \left(\sum_{i=1}^n x_i^2 \right)^{1/2};$$

$x \geq y$ denotes that $x_i \geq y_i$ $(i = 1, \ldots, n)$, where $x := [x_1, \ldots, x_n]^T$, $y := [y_1, \ldots, y_n]^T$;

$\mathbb{E}_n^+ := \{ x \in \mathbb{E}_n \mid x \geq 0 \}$;

$(x, y) := \sum_{i=1}^n x_i y_i$ —is the scalar product of the vectors x and y;

$\alpha^+ := \max\{0, \alpha\}$, where α is a number;

$x^+ := [x_1^+, \ldots, x_n^+]^T$;

$\|x\|_0 := \max_{(i)} |x_i|$; $\quad \|x\|_1 := \sum_{i=1}^n |x_i|$;

$|x - M| := \inf\{\|x - y\| \mid y \in M\}$, where $M \subset \mathbb{R}^n$;

$|x|_{\max} := \max_{(i)} x_i$, where $x^T := [x_1, \ldots, x_n]$;

$|x|_{\min} := \min_{(i)} x_i$, where $x^T := [x_1, \ldots, x_n]$;

$\dfrac{\partial f(x)}{\partial l}$ —is the derivative of $f(x)$ in the point $x \in \mathbb{E}_n$ in the direction $l \in \mathbb{E}_n$, i.e.,

$$\frac{\partial f(x)}{\partial l} := \lim_{t \to +0} \frac{f(x + tl) - f(x)}{t};$$

If P is a certain optimization problem, then $\operatorname{Arg} P$ is its optimal set; $\arg P$ is an element from $\operatorname{Arg} P$; $\operatorname{opt} P$ is the optimal value of the problem P.

If Q is the problem $\max_{x \in X} \min_{u \in Y} F(x, u)$,
then $\operatorname{Arg}_x Q := \operatorname{Arg} \max_{x \in X} \varphi(x)$ where $\varphi(x) := \min_{u \in Y} F(x, u)$.

lex-problem —is the lexicographic problem;

$\max_p(\cdot)$ —is the symbol of the problem of subsequent (lexicographic) optimization;

$\max_\pi(\cdot)$ —is the symbol of the problem of Pareto maximization;

Arg_π —is the symbol of Pareto set for the problem $\max_\pi(\cdot)$;

$\mathrm{dom}\, f := \{x \in D \mid f(x)$ is finite$\}$;

$\mathrm{epi}\, f := \{[x, \mu] \mid f(x) \le \mu,\, x \in \mathrm{dom}\, f\}$;

$\mathrm{co}\, M := \{\sum_{i=1}^{k} \alpha_i x^i \mid x^i \in M,\, \sum_{i=1}^{k} \alpha_i = 1,\, \alpha_i \ge 0\ (i = 1, \ldots, k)\ \forall k\}$
—a convex hull of a set $M \subset \mathbb{E}_n$;

$\mathrm{cone}\, M := \{\sum_{i=1}^{k} \alpha_i x^i \mid x^i \in M,\, \alpha_i \ge 0\ (i = 1, \ldots, k)\ \forall k\}$ —a conic hull of a set $M \subset \mathbb{E}_n$;

$\Rightarrow$ —is the implication; $\Leftrightarrow$ —is the equivalence;

$\exists$ —existential quantifier; $\forall$ —universal quantifier; $\forall j \in S$ for all j from S;

LP —linear programming;

IP —improper problem;

IP LP —improper problem of linear programming.

Bibliography

Alekseev, V. M., Galeev, E. M., and Tikhomirov, V. M. (1984). *Collection of Problems in Optimization. Theory. Examples. Problems.* Nauka, Moscow (in Russian).

Arrow, K. J., Hurwicz, L., and Uzawa, H. (with contributions by Chenery, H. B., Johnson, S. M., Karlin, S., Marschak, T., and Solow, R. M.) (1958). *Studies in Linear and Nonlinear Programming.* Stanford University Press, Stanford, California.

Ashmanov, S. A. (1981). *Linear Programming.* Nauka, Moscow (in Russian).

Ashmanov, S. A., and Timokhov, A. V. (1991). *Optimization Theory in Problems and Exercises. (Classical Fields. Linear Programming. Convex Sets.)* Nauka, Moscow (in Russian).

Astaf'ev, N. N. (1982). *Linear Inequalities and Convexity.* Nauka, Moscow (in Russian).

Bazaraa, M. S., and Shetty, C. M. (1979). *Nonlinear Programming. Theory and Algorithms.* John Wiley & Sons, New York.

Belen'kii, V. Z., Volkonskii, V. A., Ivankov, S. A., Pomanskii, A. B., and Shapiro, A. D. (1974). *Iterative Methods in Game Theory and Programming.* Nauka, Moscow (in Russian).

Belousov, E. G., and Andronov, V. G. (1993). *Solvability and Stability of Problems of Polynomial Programming.* Moscow State University, Moscow (in Russian).

Bel'tjukov, B. A., and Bulatov, V. P. (Eds.) (1987). *Methods of Numerical Analysis and Optimization*. Nauka, Novosibirsk (in Russian).

Benchekroun, B., and Falk, J. E. (1991). A nonconvex piecewise linear optimization problem. *Comput. Math. Appl.*, 21 (6-7), 77–85.

Braithwaite, R. B. (1959). A terminating iterative algorithm for solving certain games and related sets of linear equations. *Naval Res. Logist. Quart.*, 6 (1), 63–74.

Brown, G. W. (1951). Iterative solution of games by fictitious play. In: *Activity Analysis of Production and Allocation* (Ed. T. C. Koopmans). John Wiley & Sons, Inc., New York; Chapman & Hall, Ltd., London, 374–376.

Bulavskii, V. A., Zvjagina, R. A., and Jakovleva, M. A. (1977). *Numerical Methods of Linear Programming (Special Problems)*. Nauka, Moscow (in Russian).

Chernikov, S. N. (1968). *Linear Inequalities*. Nauka, Moscow (in Russian).

Céa, J. (1971). *Optimization: Théorie et Algorithmes*. Dunod, Paris.

Clarke, F. H. (1983). *Optimization and Nonsmooth Analysis*. John Wiley & Sons, New York—Chichester.

Dantzig, G. B. (1963). *Linear Programming and Extensions*. Princeton University Press, Princeton, New Jersey.

Davydov, E. G. (1990). *Operation Research*. Vysshaya Shkola, Moscow, (in Russian).

Dem'janov, V. F., and Vasil'ev, L. V. (1981). *Nondifferentiable Optimization*. Nauka, Moscow (in Russian).

Ekeland, I., and Temam, R. (1976). *Convex Analysis and Variational Problems*. North-Holland Publ. Co., Amsterdam—Oxford; American Elsevier Publ. Co., Inc., New York.

Elster, K.-H. (Ed.) (1993). *Modern Mathematical Methods of Optimization*. Academie Verlag, Berlin.

Elster, K.-H., Reinhardt, R., Schäuble, M., and Donath, G. (1977). *Einführung in die Nichtlineare Optimierung*. BSB B.G. Teubner, Leipzig.

Eremin, I. I. (1988). *Inconsistent Models of Optimal Planning*. Nauka, Moscow (in Russian).

Eremin, I. I. (1996). On the penalty method in mathematical programming. *Dokl. Ross. Akad. Nauk*, 346 (4), 459–461 (in Russian).

Eremin, I. I., and Astaf'ev, N. N. (1976). *Introduction to the Theory of Linear and Convex Programming*. Nauka, Moscow (in Russian).

Eremin, I. I., and Mazurov, V. D. (1979). *Nonstationary Processes of Mathematical Programming*. Nauka, Moscow (in Russian).

Eremin, I. I., Mazurov, V. D., and Astaf'ev, N. N. (1983). *Improper Problems of Linear and Convex Programming*. Nauka, Moscow (in Russian).

Ermol'ev, Yu. M., Lyashko, I. I., Mikhalevich, V. S., and Tyuptya, V. I. (1979). *Mathematical Methods of Operation Research*. Vischa Shkola, Kiev (in Russian).

Evtushenko, Yu. G. (1982). *Methods of Solution of Extremal Problems and Their Application in Optimization Systems*. Nauka, Moscow (in Russian).

Fedorov, V. V. (1979). *Numerical Methods of Maximin*. Nauka, Moscow (in Russian).

Fiacco, A. V., and McCormick, G. P. (1968). *Nonlinear Programming: Sequential Unconstrained Minimization Techniques*. John Wiley & Sons, Inc., New York.

Frolov, V. N. (1986). *Optimization of Plan Programs Under Weakly Coordinated Constraints*. Nauka, Moscow (in Russian).

Gabasov, R., and Kirillova, F. M. (1981). *Methods of Optimization*. Byelorussia State University, Minsk (in Russian).

Gale, D. (1960). *The Theory of Linear Economic Models*. McGraw-Hill Book Co., Inc., New York—Toronto—London.

Gass, S. I. (1958). *Linear Programming. Methods and Applications*. McGraw-Hill Book Co., Inc., New York—Toronto—London.

Germejer, Yu. B. (1971). *Introduction to the Theory of Operation Research*. Nauka, Moscow (in Russian).

Gill, P. E., and Murrey, W. (Eds.) (1974). *Numerical Methods for Constrained Optimization.* Academic Press, London.

Gill, P. E., Murrey, W., and Wright, M. H. (1981). *Practical Optimization.* Academic Press (Harcourt Brace Jovanovich, Publ.), London—New York.

Gol'stein, E. G. (1971). *Duality Theory in Mathematical Programming and Its Applications.* Nauka, Moscow (in Russian).

Gol'stein, E. G., et al. (1991). *Optimization Methods in Economathematical Modelling.* Nauka, Moscow (in Russian).

Gol'stein, E. G., and Tret'jakov, N. V. (1989). *Modified Lagrange Functions. Theory and Methods of Optimization.* Nauka, Moscow (in Russian).

Gol'stein, E. G., and Yudin, D. B. (1966). *New Directions in Linear Programming.* Sovetskoe Radio, Moscow (in Russian).

Gorokhovik, V. V. (1990). *Convex and Nonsmooth Problems of Vector Optimization.* Navuka i Tekhnika, Minsk (in Russian).

Gorokhovik, V. V., and Zor'ko, O. I. (1994). Piecewise affine functions and polyhedral sets. *Optimization,* 31 (3), 209–221.

Grossman, K., and Kaplan, A. A. (1981). *Nonlinear Programming on the Basis of Unconstrained Optimization.* Nauka, Novosibirsk, (in Russian).

Halmos, P. R. (1958). *Finite-Dimensional Vector Spaces.* D. Van Nostrand Co., Inc., Princeton, N.J.

Himmelblau, D. M. (1972). *Applied Nonlinear Programming.* McGraw-Hill Book Co., New York.

Ioffe, A. D., and Tikhomirov, V. M. (1974). *The Theory of Extremal Problems.* Nauka, Moscow (in Russian).

Ivanov, V. K., Vasin, V. V., and Tanana, V. P. (1978). *Theory of Linear Ill-Posed Problems and Its Applications.* Nauka, Moscow (in Russian).

Kantorovich, L. V. (1939). *Mathematical Methods of Organization and Planning of Production.* Leningrad State University, Leningrad (in Russian).

Kantorovich, L. V., and Gorstko, A. B. (1972). *Optimal Solutions in Economics.* Nauka, Moscow (in Russian).

Karlin, S. (1959). *Mathematical Methods and Theory in Games, Programming and Economics. Vol. I: Matrix Games, Programming, and Mathematical Economics. Vol. II: The Theory of Infinite Games.* Addison-Wesley Publ. Co., Inc., Reading, Mass.—London.

Karmanov, V. G. (1980). *Mathematical Programming.* Nauka, Moscow (in Russian).

Karpelevich, F. I., and Sadovskii, L. E. (1963). *Elements of Linear Algebra and Linear Programming.* Fizmatgiz, Moscow (in Russian).

Kripfganz, A., and Schulze, R. (1987). Piecewise affine functions as a difference of two convex functions. *Optimization,* 18 (1), 23–29.

Kuhn, H. W., and Tucker, A. W. (Eds.) (1956). *Linear Inequalities and Related Systems.* Princeton University Press, Princeton, New Jersey.

Künzi, H. P., and Krelle, W. (in collaboration with Oettli, W.) (1966). *Nonlinear Programming.* Blaisdell Publ. Co. (Ginn and Co.), Waltham, Mass.—Toronto, Ont.—London.

Lasdon, L. S. (1970). *Optimization Theory for Large Systems.* The Macmillan Co., New York; Collier-Macmillan Ltd., London.

Levitin, E. S. (1992). *Perturbation Theory in Mathematical Programming and Its Applications.* Nauka, Moscow (in Russian).

Lopatnikov, L. I. (1996). *Mathecon Dictionary. Dictionary of Modern Economic Science.* ABF, Moscow (in Russian).

Luce, R. D., and Raiffa, H. (1957). *Games and Decisions: Introduction and Critical Survey.* Bureau of Applied Social Research, Columbia University; John Wiley & Sons, Inc., New York.

Lyaschenko, I. N., Karagodova, E. A., Chernikova, N. V., and Shor, N. Z. (1975). *Linear and Nonlinear Programming.* Vischa Shkola, Kiev (in Russian).

Mal'tsev, A. I. (1975). *Foundations of Linear Algebra.* Nauka, Moscow (in Russian).

Mazurov, V. D. (1990). *Committee Method in Problems of Optimization and Classification.* Nauka, Moscow (in Russian).

McKinsey, J. C. C. (1952). *Introduction to the Theory of Games.* McGraw-Hill Book Co., Inc., New York—Toronto.

Melzer, D. (1986). On the expressibility of piecewise-linear continuous functions as the difference of two piecewise-linear convex functions. Quasidifferential calculus. *Mathem. Progr. Study*, No. 29, 118–134.

Minchenko, L. I., and Borisenko, O. F. (1992). *Differential Properties of Marginal Functions and Their Applications to Optimization Problems.* Navuka i Tekhnika, Minsk (in Russian).

Minoux, M. (1986). *Mathematical Programming: Theory and Algorithms.* John Wiley & Sons, Ltd., Chichester.

Moiseev, N. N., Ivanilov, Yu. P., and Stolyarova, E. M. (1978). *Methods of Optimization.* Nauka, Moscow (in Russian).

Monakhov, V. M., Belyaeva, E. S., ans Krasner, N. Ya. (1978). *Methods of Optimization. Application of Mathematical Methods in Economics.* Prosveschenie, Moscow (in Russian).

Morozov, V. A. (1987). *Methods of Regularization of Unstable Problems.* Moscow State University, Moscow (in Russian).

Morozov, V. V., Sukharev, A. G., and Fedorov, V. V. (1986). *Operation Research in Problems and Exercises.* Vysshaya Shkola, Moscow (in Russian).

Mukhacheva, E. A., and Rubinshtein, G. Sh. (1987). *Mathematical Programming.* Nauka, Novosibirsk (in Russian).

Murtagh, B. A. (1981). *Advanced Linear Programming: Computation and Practice.* McGraw-Hill Intern. Book Co., New York.

Nemchinov, V. S. (Ed.) (1959). *Application of Mathematics in Economic Investigations. Vol. 1.* Sotsekgiz, Moscow (in Russian).

Nesterov, Yu. E. (1989). *Effective Methods in Nonlinear Programming.* Radio i Svjaz', Moscow (in Russian).

von Neumann, J., and Morgenstern, O. (1953). *Theory of Games and Economic Behavior.* Princeton University Press, Princeton, N.J.

Plotnikov, S. V. (1983). *Methods of Projection in Problems of Nonlinear Programming.* Ph.D. Thesis. Ural State University, Sverdlovsk (in Russian).

Podinovskii, V. V., and Nogin, V. D. (1982). *Pareto-Optimal Solutions of Multicriteria Problems*. Nauka, Moscow (in Russian).

Polak, E. (1971). *Computational Methods in Optimization. A Unified Approach*. Academic Press, New York—London.

Polyak, B. T. (1983). *Introduction to Optimization*. Nauka, Moscow (in Russian).

Pshenichnyi, B. N., and Danilin, Yu. M. (1975). *Numerical Methods in Extremal Problems*. Nauka, Moscow (in Russian).

Ritter, K. (1969). Optimization theory in linear spaces. I. *Math. Ann.*, 182, 189–206.

Ritter, K. (1969). Optimization theory in linear spaces. II. On systems of linear operator inequalities in partially ordered normed linear spaces. *Math. Ann.*, 183, 169–180.

Ritter, K. (1970). Optimization theory in linear spaces. III. Mathematical programming in partially ordered Banach spaces. *Math. Ann.*, 184, 133–154.

Robinson, J. (1951). An iterative method of solving a game. *Annals of Math. (2)*, 54, 296–301.

Rockafellar, R. T. (1970). *Convex Analysis*. Princeton University Press, Princeton, N.J.

Romanovskii, I. V. (1977). *Algorithms of Solution of Extremal Problems*. Nauka, Moscow (in Russian).

Rubinshtein, G. Sh. (1970). *Finite-Dimensional Models of Optimization. Lecture Course*. Novosibirsk State University, Novosibirsk (in Russian).

Saaty, T. L. (1968). *Mathematical Models of Arms Control and Disarmament: Application of Mathematical Structures in Politics*. John Wiley & Sons, Inc., New York—London—Sydney.

Schrijver, A. (1986). *Theory of Linear and Integer Programming*. John Wiley & Sons, Ltd., Chichester.

Sergienko, I. V., Kozeratskaya, L. N., and Lebedeva, T. T. (1995). *Investigation of Stability and Parametric Analysis of Discrete Optimization Problems*. Naukova Dumka, Kiev (in Russian).

Shatalin, S. S. (Ed.) (1989). *Economathematical Models and Methods*. Voronezh State University, Voronezh (in Russian).

Sukharev, A. G., Timokhov, A. V., and Fedorov, V. V. (1986). *A Course of Optimization Methods*. Nauka, Moscow (in Russian).

Tikhonov, A. N., and Arsenin, V. Ya. (1977). *Solutions of Ill-Posed Problems*. V. H. Winston & Sons, Washington, D.C.; John Wiley & Sons, New York—Toronto—London.

Uzdemir, A. P. (1995). *Dynamic Integer Problems of Optimization in Economics*. Izdat. Firma "Fiz.-Mat. Liter.", Moscow (in Russian).

Vajda, S. (1957). *The Theory of Games and Linear Programming*. Methuen & Co., Ltd., London; John Wiley & Sons, Ltd., New York.

Vasil'ev, F. P. (1981). *Methods for Solving Extremal Problems: Problems of Minimization in Functional Spaces, Regularization, Approximation*. Nauka, Moscow (in Russian).

Vasil'ev, F. P. (1988). *Numerical Methods for Solving Extremal Problems*. Nauka, Moscow (in Russian).

Vasin, V. V., and Ageev, A. L. (1993). *Ill-Posed Problems with A Priori Information*. Nauka, Ekaterinburg (in Russian).

Volokitin, E. P. (1979). On representation of continuous piecewise affine functions. In: *Controllable Systems. Issue 19. Discrete Extremal Problems*. Institute of Mathematics, Siberian Branch of the USSR Acad. Sci., Novosibirsk, 14–21 (in Russian).

Vorob'ev, N. N. (Ed.) (1961). *Matrix Games*. Fizmatgiz, Moscow (in Russian; the collected volume contains translations into Russian of 16 papers and a paper by Vorob'ev).

Yudin, D. B., and Gol'stein, E. G. (1969). *Linear Programming. Theory, Methods, and Applications*. Nauka, Moscow (in Russian).

Zangwill, W. I. (1969). *Nonlinear Programming. A Unified Approach*. Prentice-Hall, Inc., Englewood Cliffs, W.I.

Zukhovitskii, S. I., and Avdeeva, L. I. (1967). *Linear and Convex Programming*. Nauka, Moscow (in Russian).